Digital Forensics in the Era of Artificial Intelligence

Digital forensics plays a crucial role in identifying, analysing, and presenting cyber threats as evidence in a court of law. Artificial intelligence, particularly machine learning and deep learning, enables automation of the digital investigation process. This book provides an in-depth look at the fundamental and advanced methods in digital forensics. It also discusses how machine learning and deep learning algorithms can be used to detect and investigate cybercrimes.

This book demonstrates digital forensics and cyber-investigating techniques with real-world applications. It examines hard disk analytics and style architectures, including Master Boot Record and GUID Partition Table as part of the investigative process. It also covers cyberattack analysis in Windows, Linux, and network systems using virtual machines in real-world scenarios.

Digital Forensics in the Era of Artificial Intelligence will be helpful for those interested in digital forensics and using machine learning techniques in the investigation of cyberattacks and the detection of evidence in cybercrimes.

Digital Forensics in the Era of Artificial Intelligence

Dr. Nour Moustafa

CRC Press
Taylor & Francis Group
Boca Raton London New York

CRC Press is an imprint of the
Taylor & Francis Group, an **informa** business

First Edition published 2023
by CRC Press
6000 Broken Sound Parkway NW, Suite 300, Boca Raton, FL 33487-2742
and by CRC Press

4 Park Square, Milton Park, Abingdon, Oxon, OX14 4RN

CRC Press is an imprint of Taylor & Francis Group, LLC

ISBN: 978-1-032-24493-8 (hbk)
ISBN: 978-1-032-24468-6 (pbk)
ISBN: 978-1-003-27896-2 (ebk)

DOI: 10.1201/9781003278962

Typeset in Palatino
by codeMantra

Contents

Preface

Recently, society has relied on digitized solutions for everyday tasks, leading organizations to employ innovative solutions to power and deliver their internet-based technologies and services, including cloud, edge, and Internet of Things (IoT). Those technologies and services are susceptible to advanced persistent threats and complex attack techniques, which corrupt the resources of services and technologies, causing finical, political, and social issues. Digital forensics was coined to identify, analyze, examine, and present those malicious events, investigate their origins, and use the findings as evidence in a court of law. Artificial intelligence (AI) technology, particularly machine learning and deep learning, can identify and discover those malicious security events, automating the digital investigation process. The emergence of artificial intelligence (AI), especially machine learning and deep learning, has the potential in automating digital forensics and its investigation process. This book offers an in-depth technical description of digital forensics's fundamental and advanced techniques and tools. It also explains how machine learning algorithms can discover and investigate cybercrimes.

This book offers a technical foundation that will be practically beneficial for students, researchers, and professional digital forensics analysts and investigators. The book illustrates digital forensics and investigation techniques using machine learning and real-world applications. It examines hard disk analysis and its style architectures as part of the investigative process. It also explains the investigation of cyberattacks in Windows, Linux, and network systems and the implementation of machine learning with virtual machines in real-world scenarios. The book offers comprehensive coverage of the essential digital forensics and AI topics, including:

- Theories of digital forensics and AI technologies
- Hard disk analysis and its styles, involving Master Boot Record and GUID Partition Table
- Investigation process for Windows, Linux, and network systems
- File system examination, including FAT, NTFS, EXT, and data recovery
- Implementation of machine learning for forensics using Python and virtualization technology
- Various cases forensics case studies and practical cybercrime investigations

This book provides knowledge for both academic and professional audiences. It would also be used as a textbook for undergraduate and postgraduate academic and professional courses in Cybersecurity, Computer Science, Information Technology, and Information Science & Management. The book serves as an essential reference volume for researchers in digital forensics using AI technology. This well-timed book will build the readers' knowledge of AI-enabled digital forensics challenges and solutions. It will be helpful to practitioners, cyber-incident teams, digital forensics investigators, and analysts. To understand the theories and practices, the reader should have a working knowledge of computing fundamentals, various operating systems, Kali Linux, virtual machines, Python programming language, and a working knowledge of cybersecurity techniques and tools. The author has the background and experience in cybersecurity and AI to write this book and provide practical insights into applying AI technology for digital forensics.

Dedication and Acknowledgment

I dedicate this book to my readers for their curiosity to learn and the spirit of my parents. I enormously thank my wife, Dr. Marwa Keshk, for her divine presence that supports my creative energy to empower the new knowledge. My children (Nardeen and Moustafa) inspired me with their intuitive perspective in transforming the content for the audience. I also thank my colleagues from UNSW Canberra, DXC Technology, and other academic, industry, and defence institutions for their genuine support in my career.

Dr. Nour Moustafa

Author

Dr. **Nour Moustafa** is currently Senior Lecturer and leader of Intelligent Security Group at the School of Engineering & Information Technology, University of New South Wales (UNSW Canberra), Australia. He is also Strategic Advisor (AI-SME) at the DXC Technology, Canberra, Australia. He was a Post-doctoral Fellow at UNSW Canberra from June 2017 to December 2018. He received his Ph.D. degree in Computing from UNSW, Australia, in 2017. He obtained his Bachelor's and Master's degrees in Computer Science in 2009 and 2014, respectively, from the Faculty of Computer and Information, Helwan University, Egypt. His areas of interest include cybersecurity, particularly network security, IoT security, intrusion detection systems, statistics, deep learning, and machine learning techniques. He has several research grants from industry and defence sponsors, such as Australia's Cyber Security Cooperative Defence Centre, Australian Defence Science and Technology Group, and the Canadian Department of National Defence. He has been awarded the 2020 prestigious Australian Spitfire Memorial Defence Fellowship award. He is also a Senior IEEE Member, ACM Distinguished Speaker, and Spitfire Fellow. He has served his academic community as the guest associate editor of multiple journals, such as *IEEE Transactions on Industrial Informatics, IEEE Systems, IEEE IoT Journal, IEEE Access, Ad Hoc Networks,* and the *Journal of Parallel and Distributed Computing.* He has also served over seven leadership conferences, including as vice-chair, session chair, Technical Program Committee (TPC) member, and proceedings chair, such as the 2020–2021 IEEE TrustCom and 2020, 33rd Australasian Joint Conference on Artificial Intelligence.

Acronyms

AI	Artificial Intelligence
ANN	Artificial Neural Network
ASCII	American Standard Code for Information Interchange
CHS	Cylinder-Head-Sector
CNN	Convolutional Neural Network
DAE	Deep Auto Encoder
DDoS	Distributed Denial of Service
DEFR	Digital Evidence First Responder
DF	Digital Forensics
DL	Deep Learning
DT	Decision Tree
EXT	Extended File System
FAT	File Allocation Table
FS	File System
FTK	AccessData Forensic Toolkit
GPT	GUID Partition Table
HD	Hard Drive
HDD	Hard Disk Drive
IDS	Intrusion Detection System
IoT	Internet of Things
KNN	K-Nearest Neighbour
LBA	Logical Block Addressing
LCN	Logical Cluster Number
MBR	Master Boot Record
MFT	Master File Table
ML	Machine Learning
MLP	Multi-Layer Perceptron
NF	Network Forensics
NTFS	New Technology File System
OS	Operating System
OSI	Open System Interconnection
RNN	Recurrent Neural Network
ROC	Receiving Operating Characteristics
SSD	Solid-State Disk
SVM	Support Vector Machine
UEFI	Unified Extensible Firmware Interface
VCN	Virtual Cluster Number
VFS	Virtual File System
VM	Virtual Machine

1

An Overview of Digital Forensics

1.1 Introduction

In recent years, society's dependence on digitized solutions for everyday tasks has led organizations to utilize innovative solutions to power and deliver their internet-based services, including cloud, edge, and Internet of Things (IoT). As technology becomes an integral part of everyday life, enhancing productivity for businesses through automation, it should come as no surprise that attackers would seek to exploit these systems and the services they provide for profit. Through the internet, attacks can launch several various malicious actions such as distributed denial-of-service, scanning/probing, keylogging, malware proliferation, email spamming, click fraud, phishing, identity theft, and more [1,2]. As such, the need for reliable methods of investigation, which can be used to identify security incidents, reconstruction of events, and attribution, is evident. As a result, the discipline of digital forensics was developed.

> **Chapter objectives: This chapter is an introduction to digital forensics and its concepts. The main objectives of this chapter are as follows:**
> - To discuss the history of digital forensics and its related disciplines
> - To understand the digital forensic process
> - To learn the digital forensic investigation steps
> - To explain how artificial intelligence (AI) can be used to automate the digital forensics process
> - To discuss the various types of cybercrime and digital evidence

1.2 Practical Exercises Included in This Book

In this book, a series of practical exercises will be given to help build practical skills in digital forensics. The practise activities were designed based

DOI: 10.1201/9781003278962-1

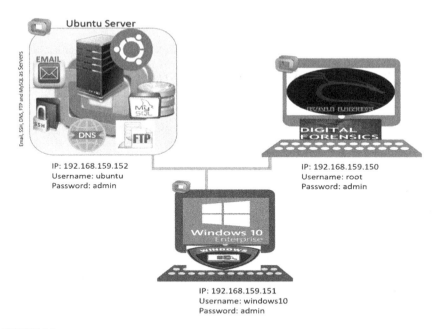

FIGURE 1.1
Virtual machines that include all exercises provided in this book.

on three virtual machines (VM),[1] an ubuntu server (IP: 192.168.159.152, username: "ubuntu," password: "admin") with several running services, a Kali (IP: 192.168.159.150, username: "root," password: "admin") equipped with several digital forensic tools and a Windows 10 (IP: 192.168.159.154, username: "windows10," password: "admin") as depicted in Figure 1.1.

1.3 A Brief History of Digital Forensics

At first, digital forensics was a misunderstood domain, with minimal space and resources being allocated for investigators to analyse digital data [3]. But as cyber threats started making their appearance, governments and law enforcement agencies began to take notice of this new discipline. One of the first conferences in the field was hosted by the Federal Bureau of Investigation (FBI) academy in 1993, "First International Conference on Computer Evidence," which gathered the attention of 26 countries. It was one of the first opportunities for practitioners from around the world to exchange ideas.

One of the first education programmes focusing on digital forensics appeared in the early 1990s when the International Association of Computer Investigative Specialists introduced training on software tools for digital forensic investigators. Gradually, various agencies started developing their digital forensic software. The internal revenue service-criminal investigation (IRS-CI) created search-warrant programmes to prepare procedures for executing warrants and search/seizure [3,4]. The first commercially available digital forensic software tool was developed by a group called ASR data. Designed for Macintosh machines, the tool was called the expert witness and allowed searches that spanned an entire hard drive [5]. Software developed and maintained by IRS, specifically the criminal investigation division, is iLook, Linux-based software allowing an investigator to acquire a complete image of a computer system and quickly review its hard drive [6].

A popular commercial product is the AccessData Forensic Toolkit (FTK) which can be used to fully examine a computer, providing email retrieval services and decrypting information found in the registry [7]. FTK is another popular digital forensic developed by Access Data. FTK's functionality specializes in scanning hard drives to retrieve deleted emails and detect strings that can be used in password dictionaries to crack encryption. It allows for forensic images of hard drives to be generated and cryptographic hashes such as MD5 and SHA1 for integrity verification.

EnCase is another popular digital forensics tool that provides threat detection and mitigation, aside from traditional forensic activities such as identification and collection. Previous modules enabled endpoint protection from data exfiltration, mitigation, and remediation of cyberattacks with minimal impact on everyday operations and provided remote worldwide forensic triage.

1.4 What Is Digital Forensics?

Locard's Exchange Principle states that any contact the criminal has had with the crime scene will leave backtraces of that interaction in the centre of all forensic disciplines and sub-disciplines [8,9]. Thus, the purpose of forensics has always been to identify these traces to assist a criminal investigation in apprehending a criminal. As criminals have expanded their activities into the internet and made use of computers to commit their crimes, law enforcement and forensics have evolved to keep up with the criminals, giving rise to a new discipline called digital forensics.

There have been multiple definitions for digital forensics, each with its own merits, proposed by different organizations and viewing the field from different perspectives [2]. One such popular purpose by McKemmish [10] is that: digital forensics is the process of identifying, preserving, analysing, and presenting digital evidence in a legally acceptable manner. As shown in Figure 1.2, based on McKemmish, the digital forensics process starts with an investigator identifying possible sources of evidence and data type. Next,

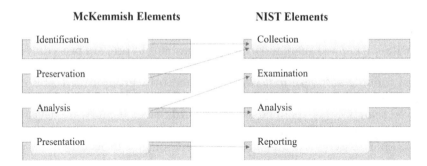

FIGURE 1.2
McKemmish and NIST elements of digital forensics.

the investigator must preserve the crime scene by ensuring that data is not altered during collection.

The collected data is processed and analysed during analysis to identify actual evidence and inferences about the case, often presented in a court of law. Evidence acquired during a digital forensic investigation may be required for many computer crimes and misuse. The collected information may be used by law enforcement, assisting in arrests and prosecution. Still, companies can also use it to terminate employment due to sabotage or misuse of corporate systems or even counter-terrorism to prevent future illegal activities.

Another definition for digital forensics, given by (National Institute of Standards and Technology) NIST [11], identifies four slightly different elements: collection, examination, analysis, and reporting. There is some overlap between the stages by McKemmish and NIST, as given in Figure 1.2. As computers rely on several specialized subsystems to function (HD, RAM, NIC, etc.), attackers may target any subsystem, depending on the purpose of the attack. Thus, in digital forensics, there are multiple methods for:

- Discovering data on computer systems
- Recovering deleted, encrypted, or damaged file information
- Monitoring live activity
- Detecting violations of corporate policy

Digital evidence can take many forms but can be thought of as any information being subject to human intervention or not that can be extracted from a system such as computers, mobile phones, and embedded devices. It must be in a human-readable format or interpreted by a person with expertise in the subject to be used in legal procedures. The digital forensic process can be seen in Figure 1.3 and is briefly discussed in Table 1.1, with more details below.

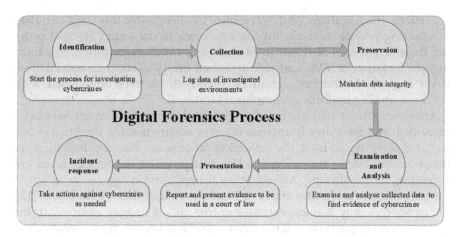

FIGURE 1.3
Digital forensic process.

TABLE 1.1

Key Digital Forensics Stages

Stages	Description
Identification	Defines the requirement for evidence management, knowing it is present, its location, and its type and format
Preservation	Concerned with ensuring evidential data remains unchanged or changed as little as possible
Analysis	Interprets and transforms the data collected into evidence
Presentation	Presents evidence to the courts in terms of providing expert testimony on the analysis of the evidence

1.4.1 Identification

Forensic experts seek to identify security events in the first stage of an investigation and determine if a crime has occurred. Investigators must establish the cybercrime type, what devices have been affected, and which need to be accessed for data collection. Identification in a cloud setting primarily focuses on establishing the requirements for carrying out the collection in a forensically sound manner. For example, identifying evidence can be done from cloud storage such as Dropbox, iCloud, and Google Drive. NIST has identified several challenges in the identification of data within a cloud instance. Furthermore, volatile forensic traces must be detected in a dynamic distributed environment.

1.4.2 Collection and Preservation

Preservation corresponds to methods and tools that ensure that extracted traces are maintained in their original form by preventing unintentional alterations, modifications, and overwrites. Challenges for this stage correlate to selecting proper methods for the assurance of data integrity and proving the effectiveness of the utilized techniques such as cryptographic hash digests. Isolation of VMs and sensitive data needs to occur in hard drives, RAM, logs, etc. In contrast, a chain of custody needs to be established, with entries spanning multiple geographical locations.

After the security incident has been detected, sources of interest have been identified, and measures have been taken to ensure that the integrity of collected data is maintained, the collection process can happen. Investigators focus on the state of the underline VM that provides the service users are accessing, by monitoring, among other things, certain logs, such as application, access, error, authentication, transaction, and data volume logs [12]. Other sources may include database entries, RAM images, network traffic (extracted or logged), and hard drives. NIST acknowledges several challenges when it comes to collecting data from the cloud: (1) extracting data from VMs, (2) ensuring data integrity in multi-tenant environments, (3) obtaining forensic images without breaching the privacy of other users, and (4) deleting data recovery from distributed virtual environments.

1.4.3 Examination and Analysis

After the data collection, the investigation process proceeds to the examination and analysis stages. In these stages, forensic investigators scan the collected data for patterns to extract initial traces that indicate a security event. The identified traces are further analysed, and inferences are extracted about the events, leading to the transformation of traces to evidence. The purpose of the investigation is to answer five questions related to a security incident, including who, what, where, when, and why.

1.4.4 Presentation

The forensic investigator will prepare a report that states the findings of the case. The report should be well-documented and include the findings from the examination and analysis, which need to be presented comprehensively to a court of law. This may include a combination of written and pictorial views. This is important because of the technical details as cloud computing is a complex environment for ordinary internet users to understand. This report will be the basis of the expert testimony on the analysis of the evidence, presenting these findings to the management of an organization to a court of law.

Presenting digital evidence is difficult and even more complicated when the investigation is in a SaaS environment. The examiner needs to be able to explain key concepts, such as (1) how the chain of custody was kept with data potentially residing on multiple devices and across different jurisdictions; time synchronization where data has been obtained from multiple time zones; (2) how they obtained legal authority to preserve and collect data which may originate in a different jurisdiction; and (3) the general complexity of cloud technologies. For criminal matters, an investigation hands the case to a prosecutor. The prosecutor has three main tasks: (1) to investigate crime, (2) to decide whether or not to instigate legal proceedings, and (3) to represent the state in court. As such, they are integral to the presentation and reporting stage. They need to convince a Judge – who is a law expert, not in particular evidence – on the merits of the case.

1.5 Artificial Intelligence for Digital Forensics

AI technology, particularly deep learning algorithms, can predict and infer cybercrime events and assist investigators in examining the origin of cyber threats. AI can be defined as a field of computer science that emphasizes addressing and solving problems that are rationally challenging for human individuals but comparatively simple for computers [2]. This includes issues that can be designated according to a group of official, arithmetic, scientific rules. AI is a comprehensive field involving learning-based methods and many other methods that do not entail any learning process. The actual problem for AI is demonstrated to be finding a solution to a problem that human individuals can efficiently perform but is challenging to be formally described by human individuals. The intuitively solved and automatic issues include attack detection, crime investigation, face recognition, object detection, etc.

In AI, machine learning (ML) is a subfield of AI, which can learn from large amounts of historical data using a set of self-adapting computational algorithms. The hard programming of computer algorithms makes it unable to cope with the dynamically changing requirements and continuously updated system conditions [13]. The efficiency of the ML algorithms is heavily reliant on selecting a proper group of features; hence, feature engineering techniques become an essential part of having a vital role in designing any ML solution. Deep learning, a subclass of ML, is inspired by the human brain to learn large amounts of data and infer hidden patterns for decision-making. Researchers focus their efforts on designing AI-based automated methods for detecting cyber incidents, such as digital forensics and intrusion detection systems, which generate alerts when anomalous or suspicious behaviour is detected [14].

In digital forensics, AI-based examination and analysis tools are being developed. It can provide investigators with interfaces that detect data structures in hard drive forensic images, extract files, emails, and messages from captured network traffic and use keywords to scan multiple files swiftly. These automated tools primarily rely on ML methods for scanning data and detecting traces of interest, utilizing models trained using supervised, unsupervised, semi-supervised, or reinforcement learning, depending on the availability of data and the intended purpose of the devised tool [15].

Digital forensics using AI is the application of scientifically proven practices and methods to identify, collect, preserve, examine, examine, and present digital traces and evidence. The need for cybercrime investigation has resulted in the formulation of digital forensics, a collection of specialized investigation methods suited for addressing the challenges of identifying, collecting, examining, and analysing evidence in the cloud for the IoT data analysed by AI technology.

1.6 Digital Forensics and Other Related Disciplines

As digital evidence can be located anywhere inside a computer, digital forensic activities vary in their complexity and application area. Examples of digital forensic activities include recovering evidence from a formatted hard drive, investigating after multiple users have taken over a system, or recovering deleted emails/files. It should be noted that digital forensics differs from data recovery. Data recovery is employed to retrieve files deleted by mistake or lost/corrupted during a power outage, while digital forensics focuses primarily on finding evidence.

Forensics investigators often work as part of a team, known as the investigation triad [16]. First, the vulnerability/threat assessment and risk management test verifies the integrity of stand-along workstations and network servers. The network intrusion detection and incident response use automated tools to detect intruders and monitors network firewall logs. Finally, digital investigations manage investigations and conduct forensics analysis of systems suspected to contain evidence.

1.7 Different Types of Digital Forensics and How They Are Used

In previous decades, before the emergence of IoT and ubiquitous computing, the sources of digital evidence were few and easily discoverable. Such sources included PCs, laptops, routers, and servers. However, the assimilation of ubiquitous computing in everyday life has caused the number of sources of evidence to increase drastically [17]. Due to the nature of IoT devices being small in size, portable, and easily concealable, forensic investigators often have a hard time identifying possible sources of evidence [17]. Thus, in the

IoT era, potential evidence sources may include cloud storage servers, client devices like Android or iOS, IoT devices like drones, smart TVs and smart metres, other cyber-physical systems, and edge computing and IoT.

1.7.1 Types of Digital Evidence

Having discussed the possible sources of evidence in the previous section, it is prudent to discuss the different types of evidence and how they may be interpreted and utilized. Through digital forensics, practically any file can be recovered, examined, and used in an investigation. Some evidence includes email, which is arguably the most important type of digital evidence, images, videos, browser search history, cell phone calls history, hard drive content, and internet communication [18].

Depending on their status, data may be categorized in one of three groups [19,20]: active data, ambient data, and archive data. Active data include files that are normally stored and maintained in the hard drive at the time of collection. In other words, these files are easily accessible, without a need to reconstruct them, although they may be otherwise protected (encrypted). Ambient data are files that have been deleted and thus exist in the hard drive, possibly fragmented. If the files were deleted from the PC, accessing them may require scanning the hard drive to be reconstructed. Finally, archive data describe data that have been stored in backup data stores. Examples of digital forensics in IoT and cloud are explained below.

1.7.1.1 Cloud Forensics in IoT

As a result of the unprecedented growth displayed by IoT and the mass adoption of this new field by organizations in areas spanning from education to industry, a novel domain emerged, the smart environment. Smart environments blend the IoT and existing business processes to improve efficiency, management, and real-time services [2]. However, this development has attracted the attention of malicious actors that, coupled with security flaws inherently found in smart sensors. This will result in new cyberattacks that target the various subsystems of IoT deployments and smart environments, seeking to exfiltrate sensitive information, disable services, exploit false measurement injections and even achieve lateral movement in otherwise secured networks.

Smart environments and IoT deployments tend to share specific architectural characteristics, although they differ significantly from vendor to vendor when implementing (technologies and protocols). Smart sensors are deployed in a room, floor, building, or collection of buildings. Each of them is responsible for monitoring some aspect of their environment (sensors) or intervening according to sensory feedback or user commands (actuators) [15]. Data collected from sensors is accumulated, processed, and forwarded to the backend servers by specialized devices called aggregators. In the backend, the data is further processed and analysed, producing analytics that, along with living feeds from specific sensors, is delivered to the end-user to provide a spherical and insightful view of the deployment. The users can then act on the information and analytics that they receive, optimizing the behaviour of the smart environment according to their preferences.

Depending on their intentions, attackers can target smart environment deployments at any part of their architecture. They can focus on exploiting weaknesses in the sensors and infect them with malware, interrupt or modify their network traffic to the server backend or access the cloud and exfiltrate information gathered from the environments. The increased attention of cyber attackers on IoT deployments has resulted in the emergence of a novel sub-discipline of digital forensics, known as IoT forensics, which can combine three sub-subdisciplines of digital forensics: hardware, network, and cloud forensics.

The investigation of cyberattacks targeting the cloud-side of an IoT deployment may be found in virtualized environments and stored in shared physical devices with multiple tenants [14]. Cloud forensic methods are primarily applied to trace the origins of cyberattacks and find evidence that can be used in a court of law. As shown in Figure 1.4, an example of an IoT deployment in an intelligent environment is presented, where aggregators group sensors. The aggregators communicate collected data to the cloud backend, where telemetry data is processed, and analytics are produced so that users can alter the behaviour of the environment and take informed, optimized

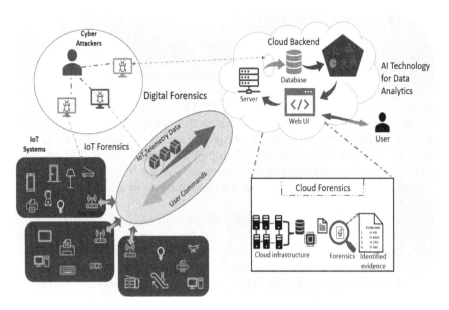

FIGURE 1.4
Cloud forensics in IoT architecture and adversary model.

decisions about how to improve efficiency in their organization. In the figure, the attackers are shown to target the IoT ecosystem in the edge layer, where sensors reside, the network, or the cloud.

1.7.1.2 Digital Forensics and Artificial Intelligence

Digital forensics and its subdomains, such as cloud forensics, define the application of scientifically proven methods for assisting law enforcement in solving crimes or security experts in identifying traces of past or ongoing cyberattacks. During an investigation, forensic experts are tasked with assessing the crime scene and identifying potential sources of evidence [2]. Next, relevant and appropriate tools are selected that are then employed to secure data extraction, establishing their integrity and a chain of custody. After the secure collection has taken place, the data, in whatever form and storage medium the investigator selected, is forwarded to a laboratory, where further actions can occur.

The forensic experts will extract a copy of the data and examine it for potential traces of abnormal activities, unusual events, and general attack-like indications. Any suspicious data, now called traces, are gathered and further analysed to answer five key questions: who was involved in the attack, what was affected, where the attack took place when the incident occurred, and what the attacker's motivations were. Answering these questions in the form of a comprehensive and concise report is one of the final responsibilities of the forensic expert.

Although this investigation process appears easy to apply, particular challenges hinder digital forensic investigations' effectiveness, timeliness, and completeness, especially in the cloud. Investigating the computer networks and systems in a crime scene involves collecting, examining, and analysing vast quantities of heterogeneous files. This process is repetitive, as the investigators need to iterate through the collected data and requires pattern detection on a level that may be beyond the capabilities of humans, which renders it error-prone. To address these challenges, researchers concentrate their efforts on designing AI-based digital forensics techniques for detecting cyber incidents that generate alerts when anomalous suspicious behaviour is detected [14]. It can provide investigators with interfaces that detect data structures in hard drive forensic images, extract files, emails, and messages from captured network traffic and use keywords to scan multiple files and define attack types and their origins.

1.8 Understanding Law Enforcement Agency Investigations

Digital forensic investigations can be carried out by either security experts working in the private sector, investigating security breaches for companies, or law enforcement agents investigating cyber-related crimes. Although, unlike investigations carried out in the private sector, law enforcement agents are bound to adhere to the laws that dictate proper investigations [21].

For example, evidence acquired in breach of the Telecommunications Interception Act 1979 [22], which prohibits access to telecommunications without the participants' knowledge, is rendered inadmissible in court. For search and seizure operations, investigators need to work within the guidelines of their warrant and either extend it or acquire a new one if necessary. Furthermore, investigators need to build a criminal case by combining all legally obtained evidence to link cybercrime to the defendant with as little doubt as possible [21].

In the United States, the Computer Fraud and Abuse Act 1986 [23] is the primary federal law covering cybercrime incidents. Since it was introduced, it has been amended several times, with an example being the use of the Patriot Act to amend the definition of "protected computers" to include any systems either situated within the United States or outside of it, which may affect commerce or communications of the United States [24]. Regardless of the scene, any forensic investigation needs to include the following objectives:

- Collect useful evidence
- Do not interrupt business processes
- Ensure that evidence has a positive impact on outcomes and legal action
- Assist any potential investigation of crimes and persuade adversaries to avoid further actions against the organization
- Offer a procedure that has an acceptable cost

1.8.1 Understanding Case Law

As stated by McKemmish [25], results produced by a digital forensic process are often used to support or prove some event or activity, often of a legal nature. Thus, the results of such an investigation must be reliable and accurate, something that can be achieved through transparency. The forensic process is verified, and confidence is built by ensuring that the entire process and results are reproducible under the same conditions. The resulting problem, thus, is as follows: how can we ensure the reliability and accuracy of the digital forensic process?

With technology changing rapidly and cybercrimes keeping pace, it becomes clear that current laws cannot keep up. Case law is used in such cases where statues either do not exist or cannot be applied [26,27]. Case law allows legal counsel to apply previous similar case rulings to current ones to address ambiguity in laws. This, however, means that examiners must be familiar with recent court rulings on search and seizure in electronic environments.

1.9 Significant Areas of Investigation for Digital Forensics

Primarily, digital forensics involves the discovery and/or recovery of data using various methods and tools available to the investigator. However, digital forensics may be applied to several diverse scenes, each requiring different methods. As such, investigations include but are not limited to [28]:

- **Data recovery** [3]: Investigating and recovering data that may have been deleted, changed to different file extensions, or hidden. It involves automated tools that access filesystems like FAT and NTFS and recover files that may have been deleted but still exist in the unallocated space of the hard drive, IoT, cloud, network systems, etc.

- **Identity theft** [12]: Many fraudulent activities ranging from stolen credit cards to fake social media profiles usually involve identity theft. The proliferation of e-banking, e-commerce, and digitizing government services (social security number) has made identity theft much more efficient. Attacks that result in identity theft include phishing and pharming.

- **Malware and ransomware investigations** [29,30]: To date, ransomware spread by Trojans and worms across networks and the internet are some of the biggest threats to companies, military organizations, organizations, and individuals. One infamous ransomware strain called CryptoLocker functioned by encrypting files in the target machine and requesting payment to decrypt them. CryptoLocker was active since 2013 and cost victims a total of $27 million by 2015. Malware can also be spread to and by mobile devices and smart devices. With malware using the IoT to spread malware, one such scenario was described by Ronen et al. [30]. The researchers showed that by hacking a Philips Hue light system and using the ZigBee protocol, a worm of their creation could infect neighbouring IoT lamps, potentially infecting an entire city.

- **Network and internet investigations** [14]: Investigating denial-of-service and distributed denial-of-service attacks and tracking down accessed devices, including printers and files. Depending on the technique used, these attacks can be categorized into volumetric, protocol, and application-based attacks. Detection methods include knowledge-based, which rely on pre-defined rules, statistical methods, which normally build a model for normal traffic (anomaly detection-based intrusion detection system) and ML with classifiers such as decision trees, artificial neural networks, and support vector machines.

- **Email investigations**: Investigating the source and IP origins, attached content, and geo-location information can all be investigated.
- **Corporate espionage**: Many companies are moving away from print copies and toward cloud and traditional disk media. As such, a digital footprint is always left behind; should sensitive information be accessed or transmitted?
- **Child pornography investigations**: Sadly, the reality is that children are widely exploited on the internet and within the deep web. With the use of technology and highly skilled forensic analysts, investigations can bring down exploitation rings by analysing internet traffic, browser history, payment transactions, email records, and images.

1.10 Following Legal Processes

For a criminal investigation to start, there needs to be a starting point. This starting point can be either evidence that someone discovered for some crime or a witness. Regardless, a witness or a victim reports a crime to the police, making an allegation. Following an allegation, police officers take statements and write a report about the crime [31]. The report is then processed by management; after that, it is decided whether to launch an investigation or log the information to a historical database where information about previous crimes is stored, called a blotter [32].

When an incident has been reported, and an investigation is launched, one of the first people to arrive at the crime scene is the digital evidence, first responder. The digital evidence first responder arrives at the scene of an incident and initially assesses the situation. Before any other action, precautions need to be taken so that no information is lost due to sudden power-offs or malicious software. A significant step that needs to be performed at this stage is establishing a chain of custody, which maintains who, where, how, why, and when interacted with the data before the investigators arrived and during the investigation [33]. Then, the digital evidence first responder acquires and preserves evidence [34].

The next stage of the investigation relies on the digital evidence specialist. The digital evidence specialist has the technical skill to analyse the collected data from the previous stage [35]. Furthermore, one of the responsibilities of the digital evidence specialist is to judge whether another specialist is to be called and assist with the investigation. Finally, to support facts about any identified evidence in a court of law, experts and witnesses are required to provide a written statement called an affidavit. An affidavit is legally binding and must include exhibits that support its claims [36].

1.11 The Cyber Kill Chain

The term "kill chain" was initially used by the military to define an enemy's steps to attack a target. In 2011, Lockheed Martin released a paper describing the "cyber kill chain" [37]. Similar in concept to the military version, this paper defines the steps used by cyber attackers in today's cyber-based attacks. Theoretically, by understanding each of these phases, defenders can better identify and stop attacks at each stage. By intercepting attackers at multiple stages of the kill chain, a defender increases their chances of shutting down the attacker's actions and denying them their objective. Since its introduction in 2011, numerous versions of the "cyber kill chain" have been proposed. This section will discuss how the human element addresses the original Lockheed Martin cyber kill chain, as depicted in Figure 1.5.

First, remember that cybersecurity awareness is nothing more than a control like encryption, passwords, firewalls, or antivirus. Security awareness is unique because it applies to and manages human risk. As security awareness addresses the human element, people often feel it does not apply to the cyber kill chain. However, that is not the case. Below are the seven stages to the

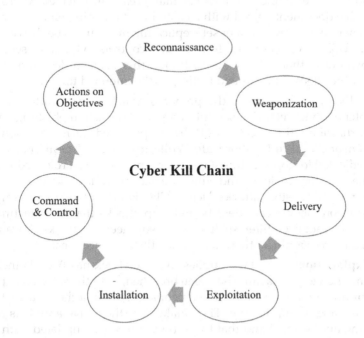

FIGURE 1.5
Cyber kill chain to understand how cyberattacks would exploit a target and follow these stages while investigating a cyber threat using digital forensics.

cyber kill chain and how a secure workforce can help neutralize a cyberattack at each of the stages [38].

- **Reconnaissance**: The attacker gathers information about the target before the actual attack is launched. Many security professionals feel that there is nothing that the defenders can do during this stage. However, that is not true. Cyber attackers often collect information on their intended targets by searching the internet sites such as LinkedIn and Instagram. In addition, they may try to gather information through techniques such as calling employees, email interactions, or dumpster diving. This is where training and discipline can have a significant impact. An aware workforce will know that they may be a target and limit what they publicly share. They will authenticate people on the phone before they share any sensitive information. They safely dispose of and shred sensitive documents. Although these actions may not entirely neutralize this stage, they can put a big dent in the attacker's capabilities.

- **Weaponization**: Having collected any necessary information from their reconnaissance, the attackers construct their attack in this stage. For example, the attacker may create an infected Microsoft Office document paired with a customized phishing email. Perhaps they make a new strain of self-replicating malware to be distributed via USB drive. There are few security controls to enhance security awareness that impact or neutralize this stage unless the cyber attacker does some limited testing on the intended target.

- **Delivery**: In this stage, the prepared attack is transmitted to the intended victim(s). For example, they send the actual phishing email or distribute the infected USB drives (prepared during weaponization) at a local coffee shop/cafe. While an entire technical industry is dedicated to stopping this stage, people play a critical role. People are the first stage to detect and eliminate many of today's attacks, including new or custom attacks such as CEO Fraud or Spear Phishing. In addition, people can identify and stop attacks that most technologies cannot even filter, such as social engineering attacks. A trained workforce significantly reduces that attack surface area.

- **Exploitation**: This stage implies that the attack has been launched and the victim machine has been breached, with the selected exploit running on the system. Trained people ensure that the systems they are using are up to date. They make sure they have antivirus running and enabled and that the antivirus has been updated with the latest attack fingerprints. An additional security measure is to make sure that any sensitive data are stored in secured systems, thus protecting them from exploits.

- **Installation**: The attacker installs malware on the victim machine. Not all attacks require malware, like a CEO fraud attack or harvesting login credentials. However, just like exploitation when malware is involved, a trained and secure workforce can help ensure they are using secure devices that are updated and have antivirus enabled, which are capable of stopping many malware installation attempts. In addition, this is where we move beyond the "human firewall" and leverage the "human sensor." An essential step in detecting an infected system is looking for abnormal behaviour. Thus, the people who interact with computer systems that may be infected are ideal for detecting abnormalities.

- **Command and control**: Once the system has been compromised, it is then managed remotely through a command-and-control infrastructure. Usually, the infected machine will initiate the communication with the command and control to signal that the infection was successful, and then the attacker can issue commands. This is the reason why "hunting," a process that involves looking for abnormal outbound activities, has become so popular.

- **Actions on objectives**: Once the cyber attacker establishes access to the target organization, they then perform actions to achieve their objectives/goal. Motivations vary depending on the threat actor but may include political, financial, or military gain, making it difficult to pinpoint those actions precisely. However, once again, this is where a trained workforce of Human Sensors embedded throughout an organization can vastly improve its ability to detect and respond to an incident. In addition, secure behaviours will make it more difficult for a successful adversary to pivot throughout the organization and achieve its objectives. Behaviours such as using strong, unique passwords, authenticating people before sharing sensitive data, or securely disposing of sensitive data are just some of the many behaviours that make the attacker's life far more difficult and result in them being far more likely to be detected.

1.12 Conclusion

In this chapter, the concept of digital forensics, its origins, and its related domains has been discussed. Furthermore, we discussed the digital forensics investigation process and legal requirements that govern how an investigation is to be conducted. Finally, we analysed the process of acquiring evidence from a cybercrime scene and discussed the cyber kill chain in detail. Next, an introduction of ML and deep learning algorithms for digital forensics will be discussed.

Note

1 https://cloudstor.aarnet.edu.au/plus/s/rwyHQRqNHuXdOhv

References

1. Bhawna Narwal, Amar Kumar Mohapatra, and Kaleem Ahmed Usmani. Towards a taxonomy of cyber threats against target applications. *Journal of Statistics and Management Systems*, 22(2):301–325, 2019.
2. Nickolaos Koroniotis, Nour Moustafa, and Elena Sitnikova. Forensics and deep learning mechanisms for botnets in Internet of Things: A survey of challenges and solutions. *IEEE Access*, 7:61764–61785, 2019.
3. Mark Pollitt. A history of digital forensics. In *IFIP International Conference on Digital Forensics*, pp. 3–15. Springer, Berlin, Heidelberg, 2010. https://doi.org/10.1007/978-3-642-15506-2_1.
4. Michael E Busing, Joshua D Null, and Karen A Forcht. Computer forensics: The modern crime fighting tool. *Journal of Computer Information Systems*, 46(2):115–119, 2005.
5. Anthony Kokocinski. Macintosh forensic analysis. In: Eoghan Casey (editor) *Handbook of Digital Forensics and Investigation*, pp. 353–382. Academic Press, Cambridge, MA, 2010.
6. Charles P Rettig and Edward M Robbins Jr. Structuring transactions and currency violations: The tax crime of the future. *Journal of Tax Practice & Procedure*, 7:35, 2005.
7. Phillip D Dixon. An overview of computer forensics. *IEEE Potentials*, 24(5):7–10, 2005.
8. Ashraf Mozayani and Carla Noziglia. *The Forensic Laboratory Handbook Procedures and Practice*. Springer Science & Business Media, Berlin, Germany, 2010.
9. Paul L Kirk. Crime investigation; physical evidence and the police laboratory, 1953.
10. Rodney McKemmish. What is forensic computing? Australian Institute of Criminology Canberra, 1999.
11. Karen Kent, Suzanne Chevalier, Tim Grance, and Hung Dang. Guide to integrating forensic techniques into incident response. *NIST Special Publication*, 10(14):800–886, 2006.
12. Lynne D Roberts, David Indermaur, and Caroline Spiranovic. Fear of cyber-identity theft and related fraudulent activity. *Psychiatry, Psychology and Law*, 20(3):315–328, 2013.
13. Nour Moustafa and Jill Slay. A network forensic scheme using correntropy-variation for attack detection. *IFIP International Conference on Digital Forensics*. Springer, Cham, India, 2018. https://doi.org/10.1007/978-3-319-99277-8_13.
14. Javed Asharf, Nour Moustafa, Hasnat Khurshid, Essam Debie, Waqas Haider, and Abdul Wahab. A review of intrusion detection systems using machine and deep learning in Internet of Things: Challenges, solutions and future directions. *Electronics*, 9(7):1177, 2020.

15. Nour Moustafa. A new distributed architecture for evaluating AI-based security systems at the edge: Network TON_IoT datasets. *Sustainable Cities and Society*, 72:102994, 2021.
16. Bill Nelson, Amelia Phillips, and Christopher Steuart. Computer forensics and investigations as a profession. In: *Guide to Computer Forensics and Investigations* (Fourth Edition), p. 5. Course Technology, Boston, MA, 2010.
17. Saad Alabdulsalam, Kevin Schaefer, Tahar Kechadi, and Nhien-An Le-Khac. Internet of Things forensics–challenges and a case study. In *IFIP International Conference on Digital Forensics*, pp. 35–48. Springer, 2018.
18. Darren Quick and Kim-Kwang Raymond Choo. Big forensic data reduction: Digital forensic images and electronic evidence. *Cluster Computing*, 19(2):723–740, 2016.
19. Jevgenijus Toldinas, Algimantas Venčkauskas, S Grigaliunas, Robertas Damaševičius, and Vacius Jusas. Suitability of the digital forensic tools for investigation of cyber crime in the Internet of Things and services. RCITD, 2015.
20. Joe Sremack. *Big Data Forensics–Learning Hadoop Investigations*. Packt Publishing Ltd, Birmingham, 2015.
21. Casey, Eoghan. *Digital Evidence and Computer Crime: Forensic Science, Computers, and the Internet*. Academic Press, Cambridge, MA, 2011.
22. Genna Churches and Monika Zalnieriute. Supplementary submission to review of the mandatory data retention regime prescribed by Part 5-1A of the Telecommunications (Interception and Access) Act 1979 (Cth)('TIA Act'). UNSW Law Research Paper 20–11, 2020.
23. Andrew Sellars. Twenty years of web scraping and the computer fraud and abuse act. *Boston University Journal of Science & Technology Law*, 24(2018):372.
24. Tiffany Curtiss. Computer fraud and abuse act enforcement: Cruel, unusual, and due for reform. *Washington Law Review*, 91:1813, 2016.
25. Rodney McKemmish. When is digital evidence forensically sound? In *IFIP International Conference on Digital Forensics*, pp. 3–15. Springer, Boston, MA, 2008. https://doi.org/10.1007/978-0-387-84927-0_1.
26. Vince Farhat, Bridget McCarthy, Richard Raysman, LLP Knight, et al. Cyber attacks: prevention and proactive responses. Practical Law, pp. 1–12, 2011.
27. Margaret A Reetz, Lauren B Prunty, Gregory S Mantych, and David J Hommel. Cyber risks: Evolving threats, emerging coverages, and ensuing case law. *Penn State Law Review*, 122:727, 2017.
28. Shiva VN Parasram. *Digital Forensics with Kali Linux: Perform Data Acquisition, Digital Investigation, and Threat Analysis Using Kali Linux Tools*. Packt Publishing Ltd, Birmingham, 2017.
29. Ronny Richardson and Max M North. Ransomware: Evolution, mitigation and prevention. *International Management Review*, 13(1):10, 2017.
30. Eyal Ronen, Adi Shamir, Achi-Or Weingarten, and Colin O'Flynn. Iot goes nuclear: Creating a zigbee chain reaction. In *2017 IEEE Symposium on Security and Privacy (SP)*, pp. 195–212. IEEE, San Jose, CA, USA, 2017.
31. Bob Morris. History of criminal investigation. In: Tim Newburn, Tom Williamson, Alan Wright (editors) *Handbook of Criminal Investigation*, pp. 41–66. Willan, Cullompton, Devon, 2012.
32. Claude Roux, Sheila Willis, and Celine Weyermann. Shifting forensic science focus from means to purpose: A path forward for the discipline? *Science & Justice*, 61(6):678–686, 2021.

33. Yudi Prayudi and Azhari Sn. Digital chain of custody: State of the art. *International Journal of Computer Applications*, 114(5):1–9, 2015.

34. Littlejohn Shinder and Michael Cross. Chapter 5: The computer investigation process. In: Littlejohn Shinder and Michael Cross (editors), *Scene of the Cybercrime* (Second Edition), pp. 201–242. Syngress, Burlington, 2008.

35. Akinola Ajijola, Pavol Zavarsky, and Ron Ruhl. A review and comparative evaluation of forensics guidelines of nist sp 800-101 rev. 1: 2014 and iso/iec 27037: 2012. In *World Congress on Internet Security (WorldCIS2014)*, pp. 66–73. IEEE, London, UK, 2014.

36. John Levingston. *The Law of Affidavits*. Federation Press, Alexandria, Australia, 2013.

37. Eric M Hutchins, Michael J Cloppert, and Rohan M Amin. Intelligence-driven computer network defense informed by analysis of adversary campaigns and intrusion kill chains. *Leading Issues in Information Warfare & Security Research*, 1(1):80, 2011.

38. Nour Moustafa, Jiankun Hu, and Jill Slay. A holistic review of network anomaly detection systems: A comprehensive survey. *Journal of Network and Computer Applications* 128:33–55, 2019.

2

An Introduction to Machine Learning and Deep Learning for Digital Forensics

2.1 Introduction

Technology is in a state of constant improvement. Creating computers with powerful processors, larger RAMs/HD, and faster networks has caused a tremendous increase in generated, benign data. However, a detrimental effect of this increase is that the sheer volume of data often obscures traces left behind by cyberattacks. This renders the process of identifying such traces difficult, as massive volumes of data need to be scanned, often in a constrained time frame. This process is impossible to perform manually by humans, and as such, automated analysis methods are necessary. To that effect, an important family of models that can scan vast quantities of data in relatively machine learning is utilized as a solution to this issue, as it can detect patterns in data that humans cannot. Deep learning is a subcategory of machine learning that has seen a lot of use in recent years. This chapter is an introduction to machine learning and its concepts.

The main objectives of this chapter are as follows:

- To understand what machine learning is
- To discuss the various types of machine learning techniques and their application
- To learn about deep learning and its capabilities
- To familiarize yourself with the process of using machine and deep learning for digital forensics
- To demonstrate a case study using a decision tree algorithm to discover attack types as the first phase in the digital forensics process using Python and a network dataset

DOI: 10.1201/9781003278962-2

2.2 History of Machine Learning

The term "machine learning" was first coined in 1952 by Arthur Samuel [1,2], who produced the first implementation of a machine learning program, that was tasked with learning to play checkers. Samuel's checker's program was unique for his time, as after each game, it "remembered" states of the game it had already seen, making changes to how it traversed its search tree in subsequent games [2].

The next big event in machine learning history was the invention of the perceptron [3] in 1958 by Frank Rosenblatt. The original perceptron was based on a hardware implementation rather than the software version that is known today, and its original design purpose was for image recognition. The idea behind the perceptron and neural networks, in general, came from studying neurons in the human brain. The idea is that mimicking the structure of clusters formed by neurons in the brain should stimulate learning. The main limitation of the original perceptron was that it could learn only linear separations in data, which rendered it unusable for learning more complex patterns.

A solution to the limitations of the first perceptron was produced by Werbos [4] with the invention of the multi-layer perceptron. As the name suggests, the multi-layer perceptron is made up of several perceptrons grouped into layers, starting from the input layer, which receives input data, the hidden layer(s) that receive the output of the input layer and process it, and the output layer. The layered architecture of multi-layer perceptrons coupled with non-linear activation functions allowed them to learn more complex patterns in data, making them very versatile.

The subsequent development in chronological order was the introduction to decision trees in 1986 by Quinlan [5]. Decision trees are directed graph-like structures comprised of nodes and edges, with each internal node representing a decision threshold that separates the data. Nodes without external edges (leaves) represent class values. Next, support vector machines (SVMs) were proposed in 1995 by Cortes and Vapnik [6]. Initially, the SVM was a binary classifier capable of linear separation, although introducing "the kernel trick" [7] can classify more complex data.

The most recent development in machine learning is a subcategory called deep learning [8]. To clarify, deep learning models are neural networks with multiple layers and/or neurons in each layer [9,10]. Deep learning has been applied to various problems, including computer vision, voice recognition, natural language processing, and more. One of the reasons why deep learning models have been the focus of attention in the research community is that, with more data, their performance increases, in contrast to older learning models.

2.3 What Is Machine Learning?

Machine learning is the discipline tasked with processing data in a structured, unstructured, or semi-structured form and drawing some inferences from it, in most cases in the form of predictions for new, unseen by the trained model, data [11]. Based on their learning approach, machine learning algorithms are separated into two main categories: supervised and unsupervised learning. A classification of machine learning and deep learning techniques used for digital forensics is shown in Figure 2.1. These techniques can be used to identify cyber threats and their origins while investigating cybercrime and/or cyber threat cases.

2.3.1 Supervised Learning

Supervised learning requires labelled data, that is, data where the class feature is known [11] [12]. Data is split into training, and testing sets, with the training set is usually larger than the testing set (often a 70% and 30% split, respectively). This allows the algorithm to make predictions during training and, based on the "distance" of the prediction from the original class feature value, also known as the training error, make corrections to the model to compensate. An appropriate loss function is selected to identify this training

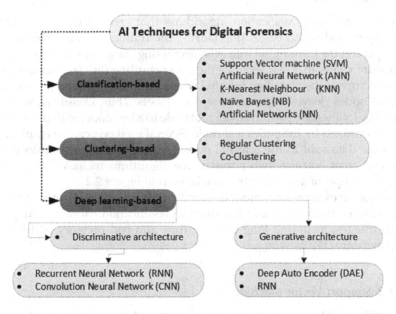

FIGURE 2.1
Popular Machine and deep learning techniques for digital forensics.

error, with the type of loss function depending on the classification problem (binary classification, multi-class classification). One example of a popular loss function, is the cross-entropy loss function (also known as log loss function) [13], as shown in Equation 2.1.

$$L = -\left(y \log\left(y^{\hat{}}\right) + (1-y) \log\left(1-y^{\hat{}}\right)\right) \qquad (2.1)$$

where "y" corresponds to the original class feature value, and "$y^{\hat{}}$" is the prediction. For binary classification, "y" takes either the value "0" or "1" (each representing one of the two classes), while "$y^{\hat{}}$" takes the values [0,1]. So, if "$y = 0$," then "$L = -(\log(1-y^{\hat{}}))$," which means that for large values of "$y^{\hat{}}$," the loss value will also be large, while for values approaching "0," the loss value will also approach 0, similarly for "$y=1$."

Supervised learning includes both classification and regression algorithms. Their main difference is that the former is used when predictions are limited to a finite set of values, while the latter is used to predict real values [12]. Classification models are tasked with learning ways to separate new data in pre-defined groups, also known as classes. Different classification algorithms view the classification problem from a different perspective, utilizing diverse approaches. Next, some classification algorithms are presented.

2.3.1.1 Decision Trees

Decision trees are classification algorithms that create tree-like structures made up of nodes and edges [14]. During the training process, the decision tree separates the training set into ever-decreasing subgroups of data by selecting feature values that best separate the data regarding the class feature. In the tree structure of the classifier, internal nodes represent decision points, while external nodes (leaves) represent the class labels. Thus, classification is performed by following a path from the root node to a leaf node, with the assigned class determined by the path's leaf node. Several methods can be employed to assess the data split in the internal nodes, such as chi-squared, information gain, and cross-entropy. Two popular tree algorithms include ID3 and C4.5 [15]. An example of a decision tree can be seen in Figure 2.2.

Decision trees are simple models, easy to understand their inner workings, with reasonable accuracy and fast during classification. However, they have disadvantages, including long training time and bad memory scaling with extensive data [14].

2.3.1.2 Support Vector Machine

Originally, SVM was developed by Cortes and Vapnik [6]. SVM models perform classification by mapping the data in a high-dimensional space. It becomes easier to construct a line that better separates the data points into

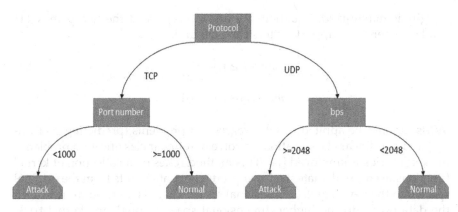

FIGURE 2.2
An example of a decision for classifying normal and attack data as the first phase of discovering cyberattacks in the digital forensics process.

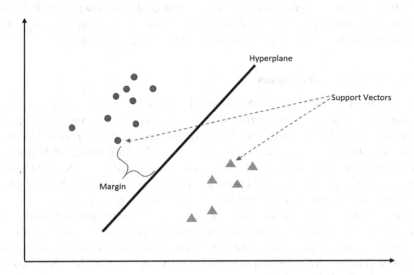

FIGURE 2.3
An example of a support vector in two dimensions.

groups. In their original form, SVMs were tasked with linear binary classification, as illustrated in Figure 2.3. A hyperplane (a space of dimension equal to one lower than that of the data space) is sought during training, separating the data points into two classes (for binary classification). It maximizes the distance between the hyperplane and the closest points called support vectors [6,14].

These support vectors form planes (often named H1 and H2) that have equal distance from the hyperplane, called margin. An SVM's task is to identify the hyperplane H, for which its distance from the support vectors, the margin, is maximized. Equations 2.2 and 2.3 represent the two planes, H1 and H2, where the support vectors can be found.

$$wx_i + b \geq 1, y_i = 1 \tag{2.2}$$

$$wx_i + b6 - 1, y_i = -1 \tag{2.3}$$

SVMs can also be applied to solve regression problems (predicting real values) and non-linear classification. To solve non-linear classification problems, the kernel trick is employed [14]. This method relies on a collection of kernel functions used to calculate the inner product of data points from the original space into the new higher-dimensional space. There is no need to transform the data points to the higher-dimensional space through the kernel trick, leading to fewer computations.

Although SVMs have some significant advantages, such as being unaffected by data dimensionality and guaranteed to find the classification function, they also have some disadvantages. Some of the disadvantages of SVMs include excessive memory requirements, time-consuming training, and computationally intensive interpretation of results is difficult [16].

2.3.1.3 K-Nearest Neighbours

The k-nearest neighbour (KNN) model belongs to a family of parametric, instance-based learning algorithms [16,17]. Such algorithms require no training instead of building their model during the classification process. Distance between data points is a crucial concept for the KNN algorithm. When a new sample is introduced to the classifier, it seeks to identify the "K" data points closest to the new data. It then classifies the new data point with the highest frequency (occurrence) out of the KNNs, as shown in Figure 2.4.

The method chosen to calculate the distance between the n-dimensional data points (where n is the number of independent features) should minimize the distance between points of the same class while maximizing distance for different classes. One example of distance measure used in the KNN classification is the Euclidian distance, given by the following equation.

$$D(a,b) = \sqrt{\sum_{i=1}^{m} (a_i - b_i)^2} \tag{2.4}$$

The number of neighbours to consider for determining the class is very important. Choosing small values can cause noise to affect the results while choosing large k values means that samples belonging to different classes may lead the model to misclassify.

KNN models include some advantages, such as they are easy to understand and implement, require less time for training (no training) [16,17]. Some serious disadvantages include: they need large storage of data and extensive

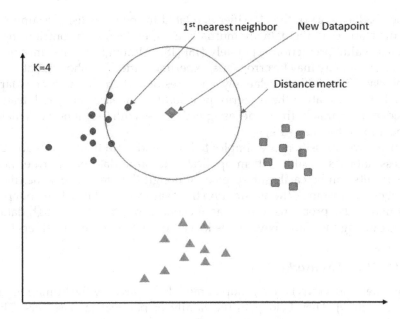

FIGURE 2.4
Example of k-nearest neighbour classification (*K*=4).

time during classification, reduced accuracy, as all features are considered to be equally as important, the number of neighbours (k) is chosen manually and heavily affects the performance of the classifier.

2.3.1.4 Naive Bayes

Naive Bayes is a group of statistical classifiers relying on probability theory to perform classification, and specifically the a posteriori probability of each class, given a sample [16]. At the time of training, estimations of the conditional probability of a class value given a sample are calculated from the training data, later used during the classification process. This conditional probability is computed by applying the Bayes' Theorem, which is given in Equation 2.5.

$$P(Y\,/\,X) = \frac{P(X\,/\,Y)P(Y)}{P(X)} \tag{2.5}$$

To understand the equation above, it can be interpreted as "given a sample *X*, the probability of that sample's class value being *Y* is equal to the conditional probability of *X* given *Y* times the probability of class *Y* occurring, divided by the occurrence of sample *X*."

The "Naive" part of the classifier is related to the unrealistic assumption that the features of the data are mutually independent, each contributing to the class value prediction separately [16] [18]. Although the assumption of mutual exclusivity may be erroneous, experimentation has shown that Naive Bayes classifiers achieve relatively good results when the dataset is large enough. By calculating the posterior probabilities for each class, each feature considered separately, the model assigns the class with the highest calculated probability to the new sample.

Naive Bayes classifiers are simpler to build and train than others and can process data fast, making them applicable to large datasets. Furthermore, their results can be easily interpreted, although they too have some disadvantages. For instance, the assumption that features in a dataset are independent may cause problems to the classification process if not enough data is given, causing the Naive Bayes to be less accurate than other classifiers [16].

2.3.1.5 Neural Networks

Neural networks (NN) are a group of models inspired by the human brain's neurons [16,19]. They belong to the family of parametric classifiers. They include parameters such as the number of neurons/layers, weights and biases, and hyperparameters, such as the number of epochs, learning rate, and batch size. A standard NN can be represented by a directed, often acyclic graph, comprised of nodes called neurons and edges (also known as synapsis), with its neurons grouped into layers. Typically, an NN has at least three layers, an input layer, hidden layer(s), and an output layer, as shown in Figure 2.5.

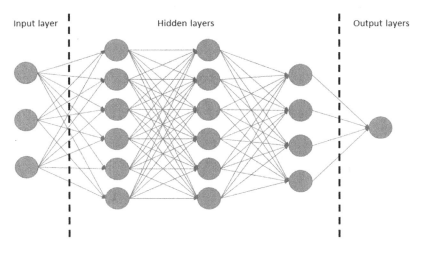

FIGURE 2.5
Example of feedforward neural network.

Each neuron in a layer received multiple inputs from the neurons of the previous layer. These inputs are weighted, summed up, and passed through an activation function, which determines whether the neuron "fired" (was activated) or not. To train the neural network, which involves updating the weights and biases of neurons in each layer, a process called backpropagation can be used [16]. During backpropagation, after the NN has processed some of the training data and calculated an estimate of the class feature value for that input, a cost function is used to estimate the neural network's training error.

The error then is "propagated" from the output layer towards the input, updating the weights and biases in the way. The update is performed by either adding or subtracting the partial derivative of the cost function, concerning weights and biases of the layer, times the learning rate. Depending on what portion of the training data is processed by the NN before the weights and biases are updated, there exist three main strategies:

- **Batch gradient descent**: Process the entire training set before updating the network's weights. Slow training usually requires loading the whole training to RAM (ineffective) and gets stuck in local minima.

- **Mini-batch gradient descent**: Process a subset of the entire training set, called a batch, before updating the network's weights, which is faster than batch gradient descent.

- **Stochastic gradient descent**: Process one record from the training set before updating the network's weights and repeating for the entire training set. Quicker updates to weights but fluctuates wildly, which may require excess time to converge to the minima.

Because of their capabilities, multiple NN subtypes have been proposed over the years, each differing in its implementation and strengths and best suited for different tasks. For example, convolutional neural networks (CNN) [20] were developed for image processing and are widely used for computer vision, while recurrent neural networks (RNN) [21] record sequential data and are often used for natural language processing, speech recognition, machine translation, and more.

2.3.2 Unsupervised Learning

Unsupervised learning algorithms function by receiving unlabelled data. The class feature is unknown and is tasked with identifying the underlying patterns in data [22,23]. For example, in clustering, which is one type of unsupervised learning, data is processed by a clustering algorithm that uses some measure of similarity to gauge whether a data point belongs in a group of similar data points, also known as clusters. This similarity measure is often expressed as a distance between records (data points), with the features being the records dimensions. One prominent example of distance measure is the Euclidean distance between two points in the k's dimensional space, as given in Equation 2.6.

$$d(x,y) = \sqrt{\sum_{i=1}^{k} (x_i - y_i)^2} \; k \in N^+ \tag{2.6}$$

Many such clustering algorithms have been developed over the years, with one of the better-known ones being the K-Means algorithm. In the K-Means algorithm, a user-defined "K" indicates the number of clusters, and membership to a cluster is established by calculating the distance of the new data point to all the centroids of the available clusters. The algorithm initially uses random centroids (which represent the centre of a cluster). Iteratively, by re-calculating distances and adjusting the clusters, new centroids are calculated until either the number of iterations has been reached or the centroids remain relatively unchanged between iterations.

2.4 What Is Deep Learning

Deep learning is a subcategory of artificial neural networks (ANNs) and thus a subset of machine learning and artificial intelligence. Like ANNs, deep learning models use "neurons" that produce an output that indicates whether the neuron is active or inactive based on their input and an activation function. Additionally, deep learning incorporates both supervised and unsupervised models. The main difference from normal ANNs is that deep neural networks often have a deep architecture comprised of many neurons and layers [10]. Although there are no strict rules to distinguish between a shallow and a deep ANN architecture, the former often have one to three layers, while the latter may have hundreds.

One way to view the benefits of a deep neural network's architecture is to consider that as the information propagates from the input towards the network's output, the layers process higher-level data features increasingly. For instance, in a CNN that is used to process an image and detect objects in it, the level of features progressively increases with the network's layer depth (as we move towards the output). For example, suppose the CNN is tasked with identifying a car in an image. In that case, initial layers may recognize lines and curves in the image. Later layers identify parts of the vehicle (tires, lights, etc.) until the network has identified the vehicle. Deep learning models are separated into discriminative and generative models [24], as explained below.

2.4.1 Discriminative Deep Learning

Discriminative are the models tasked with classification (supervised learning), which focus on separating the data by calculating the conditional

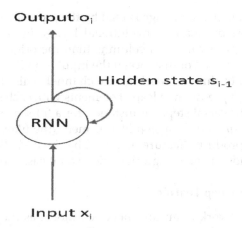

FIGURE 2.6
Example of recurrent neural network.

probability of the class, considering the input data ($P(Y/X)$). Some prominent examples of discriminative deep learning models are explained below.

2.4.1.1 Recurrent Neural Network (RNN)

An RNN can be useful as a discriminative model when the information maintains some temporal relation to its previous states. As shown in Figure 2.6, RNNs have three main components, an input node, where the data is introduced to the network, a hidden state, where the input is combined with the hidden state of the previous iteration. Thus, "memory" in the calculation of the new hidden state is used in RNN, and the output layer produces the network's output.

Due to their ability to use memory, RNNs have been applied to various problems, including machine translation, speech recognition, image description generation, image and music composition, and more. Although versatile, the way that RNNs use previous states in their hidden state calculations has been proven to have some limitations for extended sequences of data. As a result, an improved version of the RNN named the long short-term memory was developed. Their main difference with RNNs is that the long short-term memory is equipped with a different mechanism that determines when the memory of the network is updated, thus somewhat regulating its memory [25].

2.4.1.2 Convolutional Neural Network (CNN)

A CNN is another deep neural network often employed to visualize data (images, videos, etc.). They are a space-invariant neural network inspired by the connections neurons form in the brain's visual cortex. A CNN architecture consists of an input layer, an output layer, and several hidden layers such as convolutional layers, pooling layers, fully connected layers, and normalization layers.

In a CNN architecture, an image is read by the input layer and forwarded to the first of possibly many convolutional layers. In each convolutional layer, a filter has fewer rows and columns than the original image but the same depth as the input. For instance, if the input is an RGB image, the filter will have a depth of three for each image channel. This traverses through the input, making a pre-defined leap of elements with each step, extracting a number for each traversal step. For instance, for a filter $n*n*3$ and an image of $k*k*3$, where $k>n$, and for a step of 1 element (the filter shifts 1 element to the right), the produced feature map will be $(k-n+1)*(k-n+1)$. Popular applications include image recognition, video analysis, and NLP [26].

2.4.2 Generative Deep Learning

Generative models work in an unsupervised mode, as they do not require labelled data and produce the joint probability of observation and the class $(P(X, Y))$. Some prominent generative model examples are deep auto-encoder and RNN, explained below.

2.4.2.1 Deep Auto Encoder

Deep autoencoders are types of ANN tasked with learning efficient representations of the features in a dataset [27]. It comprises an input layer, a hidden layer that is often smaller than the input in neurons, and an output layer with the same dimensions as the input. The input and hidden layers are called an encoder, as they convert the input into a different representation (fewer features, more efficient, compressed).

In contrast, the hidden and output layers are called a decoder, as this part of the network is tasked with converting the "coded" features back into their original form. Deep autoencoders work in an unsupervised mode, as they are tasked with recreating the input with as few errors as possible. They are primarily employed to reduce the dimensionality of data by learning an efficient "code" of the features by stacking the encoder to the input of another classifier.

2.4.2.2 Recurrent Neural Network (RNN)

RRNs can function in both supervised (as previously mentioned) and unsupervised modes. When used in unsupervised learning, RNNs can be taught sequences of musical notes, text, and images, and after they have been trained, they can automatically generate new data [28,29].

2.5 Evaluation Criteria of Machine and Deep Learning

Machine learning and deep learning algorithms introduced for the Internet-of-Things security (attack detection) often make use of the following

		Actual	
		Normal	**Attack**
Predicted	**Normal**	True Positive (TP)	False Negative (FN)
	Attack	False Positive (FP)	True Negative (TN)

FIGURE 2.7
Confusion matrix to estimate the correct and incorrect examples of the machine and deep learning algorithms.

criteria in the evaluation purposes, namely accuracy, precision, recall, *F*1-score, true positive rate, false alarm rate, false positive rate, receiver operating characteristic curve, and area under the curve (AUC). Most of these performance criteria can be estimated using a confusion matrix [30].

It is a matrix-like representation of the classification outcomes. True positive and true negative represent the number of cyberattacks and legitimate instances correctly classified by the algorithms' outputs. False positive and false negative denote the number of legitimate and cyberattack records erroneously classified by the algorithms. The structure of the confusion matrix in the intrusion detection challenge as the first stage in digital forensics is depicted in Figure 2.7.

Accuracy: It estimates the rate of correctly classified data instances by the algorithms to the total number of instances. The accuracy is computed by:

$$\text{Accuracy } (A) = \frac{TP + TN}{TP + TN + FP + FN} \tag{2.7}$$

Precision: It computes the rate of instances rightly classified by the algorithms to the total number of predicted cases. The accuracy is estimated by:

$$\text{Percision } (P) = \frac{TP}{TP + FP} \tag{2.8}$$

Recall: It represents the proportion of samples correctly categorized by the model to the total number of actual class samples.

$$\text{Recall}(R) = \frac{TP}{TP + FN} \tag{2.9}$$

F1-measure: It is a harmonic average of the precision and recall of the algorithms' outputs. It considers false positive and false negative rates estimated by the confusion matrix. It is computed as given:

$$F1 - \text{measure} = 2 \times \frac{P \times R}{P + R} = \frac{2TP}{2TP + FP + FN} \tag{2.10}$$

There are circumstances for which assigning a rate more significance to either recall or precision is demanded. It needs modifying the above formula a little such that it can incorporate a variable parameter β beta to fulfil this target. This updated F1-measure can be generalized for various classification dilemmas using beta-version $F1 - \text{measure}_\beta$, which is estimated by:

$$F1 - \text{measure}_\beta = \left(1 - \beta^2\right) \times \frac{P \times R}{\left(\beta^2 \times P\right) + R} \qquad (2.11)$$

A β-parameter allows managing the trade-off of significance amongst the recall and precision measures. More focus is on precision when $\beta < 1$ while more emphasis is on the recall when $\beta > 1$.

True positive rate (TPR): It computes the proportion of instances correctly categorized belonging to a given class to the total number of cases in this class. TPR is estimated by:

$$\text{TPR} = \frac{\text{TP}}{\text{TP} + \text{FN}} \qquad (2.12)$$

False positive rate (FPR): It computes the proportion of instances incorrectly defined belonging to a given class to the total number of instances in this class. FPR is calculated by:

$$\text{FPR} = \frac{\text{FP}}{\text{TN} + \text{FP}} \qquad (2.13)$$

The receiving operating characteristic curve represents the relation of true positive rate and false positive rate, as shown in Figure 2.8. Each data point in receiving operating characteristics offers the performance of the machine and deep learning algorithms on a specific distribution. Following the outcomes of receiving operating characteristics, the curve is a suitable performance measure that estimates the scale of separability, which examines whether the algorithms can successfully disseminate between various cyberattack classes.

2.6 Case Study of Machine Learning-Based Digital Forensics

In this case study, a network data source, such as UNSW-NB15 [31] and TON_IoT [32] datasets, can be used to discover cyberattacks by applying a machine/deep learning algorithm. Python will be used to demonstrate to execute an attack detection as the first stage in Digital forensics to discover attack types in a data set.

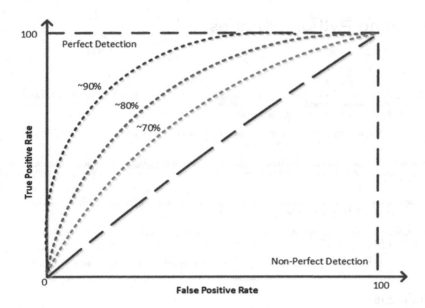

FIGURE 2.8
The receiving operating characteristic curve to represent the performance of the machine and deep learning algorithms.

FIGURE 2.9
Running Jupyter notebook.

- To do so, first, run the Jupyter notebook on the Mobile System virtual machine, as shown in Figure 2.9.
- As shown in Figure 2.10, load the following Python libraries to enable processing data and execute the machine learning algorithms for classifying attack types.
- Read the dataset using the Pandas library in the Jupyter notebook, as shown in Figure 2.11.
- It is essential to transform the data into numeric formats, which enhances the performance of the machine and deep learning models. To do this, there are various feature conversion or transformation methods such as label encoder and one hot encoding ones, as shown in Figure 2.12.

The python code of applying the label encoder and one-hot encoding is presented in Figure 2.13.

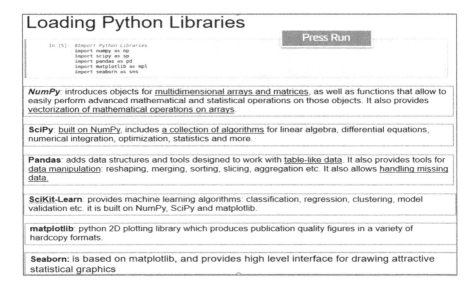

Loading Python Libraries

```
In [5]: #Import Python Libraries
        import numpy as np
        import scipy as sp
        import pandas as pd
        import matplotlib as mpl
        import seaborn as sns
```

Press Run

NumPy: introduces objects for <u>multidimensional arrays and matrices</u>, as well as functions that allow to easily perform advanced mathematical and statistical operations on those objects. It also provides <u>vectorization of mathematical operations on arrays</u>.

SciPy: <u>built on NumPy</u>, includes <u>a collection of algorithms</u> for linear algebra, differential equations, numerical integration, optimization, statistics and more.

Pandas: adds data structures and tools designed to work with <u>table-like data</u>. It also provides tools for <u>data manipulation</u>: reshaping, merging, sorting, slicing, aggregation etc. It also allows <u>handling missing data.</u>

SciKit-Learn: provides machine learning algorithms: classification, regression, clustering, model validation etc. it is built on NumPy, SciPy and matplotlib.

matplotlib: python 2D plotting library which produces publication quality figures in a variety of hardcopy formats.

Seaborn: is based on matplotlib, and provides high level interface for drawing attractive statistical graphics

FIGURE 2.10
Python libraries needed for executing machine learning.

```
#Read csv file
df = pd.read_csv("/home/admin1/Desktop/ML/network_data.csv")
print (df)
```

	proto	saddr	daddr	state	dttl
0	arp	192.168.159.150	192.168.159.151	INT	0
1	udp	192.168.159.1	192.168.159.254	CON	16

FIGURE 2.11
Reading data using Pandas.

- The decision tree algorithm is used in this case study to classify attack data and demonstrate how it assists the investigators in starting the digital forensics process. And then, the investigator can follow the legal process and find digital evidence (e.g., the attack types and their origin). This evidence can be presented in a report that can be understood in a court of law. The entire Python code of using the decision tree and its outputs are shown in Figure 2.14.

The Python code above illustrates using a small sample network data chosen to classify attack types. Then, a conversion method was used to enhance the performances, such as accuracy, false-positive rates, and false-negative rates of the decision tree algorithm. The outputs of this algorithm were explained using the confusion matrix and accuracy. It can be seen that the model achieved (a 0.97 accuracy – 97% using small network data portions). This does not mean the model performs very well.

- **Feature conversion methods** transform categorical values into numeric ones.
  - **Label Encoder:** It is used to transform non-numerical labels to numerical labels (or nominal categorical variables).
    - ➢ Numerical labels are always between 0 and n classes-1.
    - ➢ For example, the protocol attribute (proto) includes ordinal values **(TCP, UDP, FTP)** that are transformed into **(0,1,2),** respectively.
    - ➢ it may <u>decrease performance of a model</u>, such as proto has numeric values that will be like other attributes such as states. So, the model can not discriminate between proto and state attributes.
  - **One Hot Encoding (Dummy Coding):** is a commonly used method for converting a categorical input variables into continuous variables.
    - ➢ 'Dummy', as the name suggests is a duplicate variable which represents one level of a categorical variable. Presence of a level is represented by 1 and absence is represented by 0. Its challenge is <u>increasing number of features (high dimensions).</u>
    - ➢ For example, **proto (TCP, UDP, FTP)** is converted into a binary table with three columns:

Proto	Proto_TCP	Proto_UDP	Proto_FTP
TCP	1	0	0
UDP	0	1	0

FIGURE 2.12
Popular feature conversion methods.

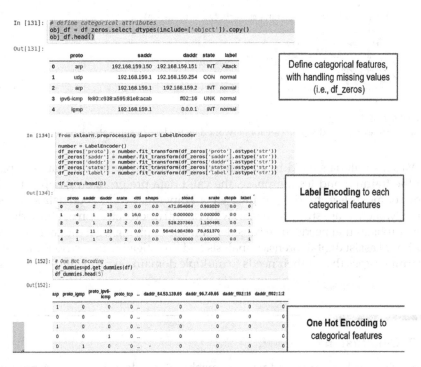

FIGURE 2.13
Examples of label encoding and one hot encoding methods.

```
 8 import numpy as np
 9 import pandas as pd
10 from sklearn.model_selection import train_test_split
11 from sklearn.metrics import accuracy_score , confusion_matrix
12 from sklearn . tree import DecisionTreeClassifier
13
14 df = pd. read_csv ('network_data_numbers.csv')
15
16 print (df. head (3))
17
18 y = df['label']. values
19 print ('Y Shape', y. shape)
20
21 X = df. drop ('label', axis =1) . values
22 print ('X Shape', X. shape)
23
24 # divide data into training and testing sets
25 X_train , X_test , y_train , y_test = train_test_split (X, y, test_size =0.2, random_state =1)
26
27 # select model
28 model = DecisionTreeClassifier ()
29
30 # train the model
31 model.fit ( X_train , y_train )
32
33 # prediction using the testing phase
34 y_pred = model.predict ( X_test )
35
36 # Measuring performance using Accuracy
37 print ("Accuracy", accuracy_score (y_pred , y_test ))
38
39 #Measuring the performance using Confusion Matrix
40 print ("Confusion Matrix", confusion_matrix (y_pred , y_test ))
```

```
In [1]: runfile('/home/admin1/Desktop/ML/Decision tree.py', wdir='/home/
admin1/Desktop/ML')
     dttl  shops      sload    srate  dtcpb  label
0     0       0   471.854004  0.983029      0      1
1     16      0     0.000000  0.000000      0      0
2     0       0   528.237366  1.100495      0      0
Y Shape (2000,)
X Shape (2000, 5)
Accuracy 0.97
Confusion Matrix [[101   11]
 [  1 287]]
```

FIGURE 2.14
Example of executing decision tree for network data to classify some attack records.

Various criteria should be considered, such as the number of data samples in the training and testing phase, the valid data pre-processing applied, and environments (e.g., hard drive, Internet of Things, network, cloud data sources). Finally, such a case study resonates with the intrusion detection or attack detection problem that needs to be considered while applying legal digital forensics. This will assist digital forensics analysts and investigators use various machine learning types that fit their needs to multiple domains explained in Chapter 1.

2.7 Conclusion

This chapter has explained various machine and deep learning techniques for digital forensics. It explained types of machine learning, including supervised and unsupervised methods. Then, it also demonstrated deep learning and its main types of discriminative and generative architectures. Moreover, it discusses the most common evaluation criteria that can assess the performances of a machine and deep learning models. Then, a practical case study was then explained to illustrate how machine learning could be applied to digital forensics challenges such as intrusion and attack detection. Next, forensics and computer foundations will be explored to demonstrate how digital forensics will examine hard disks.

References

1. Ron Kohavi. Glossary of terms. *Special Issue on Applications of Machine Learning and the Knowledge Discovery Process*, 30(271):127–132, 1998.
2. Arthur L Samuel. Some studies in machine learning using the game of checkers. II-recent progress. *IBM Journal of Research and Development*, 11(6):601–617, 1967.
3. Frank Rosenblatt. The Perceptron, a perceiving and recognizing automaton Project Para. Cornell Aeronautical Laboratory, 1957.
4. Paul J Werbos. Beyond regression: New tools for prediction and analysis in the behavioral sciences. Ph. D. thesis, Harvard University, Cambridge, MA, 1974.
5. J Ross Quinlan. Induction of decision trees. *Machine Learning*, 1(1):81–106, 1986.
6. Corinna Cortes and Vladimir Vapnik. Support-vector networks. *Machine Learning*, 237–297:20, 1995.
7. Martin Hofmann. Support vector machines-kernels and the kernel trick. *Notes*, 26:1–16, 2006.
8. Geoffrey E Hinton. Learning multiple layers of representation. *Trends in Cognitive Sciences*, 11(10):428–434, 2007.
9. Ian Goodfellow, Yoshua Bengio, and Aaron Courville. *Deep Learning*. MIT Press, Cambridge, MA, 2016.
10. Jürgen Schmidhuber. Deep learning in neural networks: An overview. *Neural Networks*, 61:85–117, 2015.
11. Pramod Singh. Introduction to machine learning. In: Pramod Singh (editor) *Deploy Machine Learning Models to Production*, pp. 1–54. Apress, Berkeley, CA, 2021.
12. Moustafa, Nour, Kim-Kwang Raymond Choo, and Adnan M. Abu-Mahfouz. AI-enabled threat intelligence and hunting microservices for distributed industrial IoT system. *IEEE Transactions on Industrial Informatics*, 18(3):1892–1895, 2022.
13. Khan, Izhar Ahmed, Nour Moustafa, Dechang Pi, Yasir Hussain, and Nauman Ali Khan. DFF-SC4N: A deep federated defence framework for protecting supply chain 4.0 networks. *IEEE Transactions on Industrial Informatics*, 2021. doi: 10.1109/TII.2021.3108811.
14. Tristan Fletcher. Support vector machines explained. Tutorial paper, 2009.
15. Badr Hssina, Abdelkarim Merbouha, Hanane Ezzikouri, and Mohammed Erritali. A comparative study of decision tree id3 and c4. 5. *International Journal of Advanced Computer Science and Applications*, 4(2):13–17, 2014.

16. Hetal Bhavsar and Amit Ganatra. A comparative study of training algorithms for supervised machine learning. *International Journal of Soft Computing and Engineering (IJSCE)*, 2(4):2231–2307, 2012.

17. Klaus Hechenbichler and Klaus Schliep. Weighted k-nearest-neighbor techniques and ordinal classification. 2004.

18. Andrew McCallum, Kamal Nigam, et al. A comparison of event models for naive bayes text classification. In *AAAI-98 Workshop on Learning for Text Categorization*, vol. 752, pp. 41–48. Citeseer, Madison, Wisconsin, United States, 1998.

19. Xavier Glorot and Yoshua Bengio. Understanding the difficulty of training deep feedforward neural networks. In *Proceedings of the Thirteenth International Conference on Artificial Intelligence and Statistics*, pp. 249–256, Sardinia, Italy, 2010.

20. Alex Krizhevsky, Ilya Sutskever, and Geoffrey E Hinton. Imagenet classification with deep convolutional neural networks. In *Advances in Neural Information Processing Systems*, pp. 1097–1105, Harrah's Lake Tahoe, United States, 2012.

21. Stephen Grossberg. Recurrent neural networks. *Scholarpedia*, 8(2):1888, 2013.

22. Rui Xu and Don Wunsch. *Clustering*, vol. 10. John Wiley & Sons, Hoboken, NJ, 2008.

23. Pang-Ning Tan, Michael Steinbach, Vipin Kumar, et al. Cluster analysis: Basic concepts and algorithms. *Introduction to Data Mining*, 8:487–568, 2006.

24. Elike Hodo, Xavier Bellekens, Andrew Hamilton, Christos Tachtatzis, and Robert Atkinson. Shallow and deep networks intrusion detection system: A taxonomy and survey. arXiv preprint arXiv:1701.02145, 2017.

25. Klaus Greff, Rupesh K Srivastava, Jan Koutník, Bas R Steunebrink, and Jürgen Schmidhuber. Lstm: A search space odyssey. *IEEE Transactions on Neural Networks and Learning Systems*, 28(10):2222–2232, 2016.

26. Maryam M Najafabadi, Flavio Villanustre, Taghi M Khoshgoftaar, Naeem Seliya, Randall Wald, and Edin Muharemagic. Deep learning applications and challenges in big data analytics. *Journal of Big Data*, 2(1):1, 2015.

27. Li Deng. A tutorial survey of architectures, algorithms, and applications for deep learning. *APSIPA Transactions on Signal and Information Processing*, 3:e2, 2014.

28. Alex Graves, Abdel-rahman Mohamed, and Geoffrey Hinton. Speech recognition with deep recurrent neural networks. In *2013 IEEE International Conference on Acoustics, Speech and Signal Processing*, pp. 6645–6649. IEEE, Vancouver, BC, Canada, 2013.

29. Peter Potash, Alexey Romanov, and Anna Rumshisky. Ghostwriter: Using an lstm for automatic rap lyric generation. In *Proceedings of the 2015 Conference on Empirical Methods in Natural Language Processing*, pp. 1919–1924, Lisbon, Portugal, 2015.

30. Nour Moustafa, Marwa Keshk, Kim-Kwang Raymond Choo, Timothy Lynar, Seyit Camtepe, and Monica Whitty. DAD: A distributed anomaly detection system using ensemble one-class statistical learning in edge networks. *Future Generation Computer Systems*, 118:240–251, 2021.

31. Nour Moustafa and Jill Slay. UNSW-NB15: A comprehensive data set for network intrusion detection systems (UNSW-NB15 network data set). *2015 Military Communications and Information Systems Conference (MilCIS)*. IEEE, Canberra, Australia, 2015.

32. Moustafa, Nour, Mohiuddin Ahmed, and Sherif Ahmed. Data analytics-enabled intrusion detection: Evaluations of ToN IoT Linux Datasets. In *2020 IEEE 19th International Conference* on Trust, Security and Privacy in Computing and Communications (TrustCom), pp. 727–735. Guangzhou, China, IEEE, 2020.

3

Digital Forensics and Computer Foundations

3.1 Introduction

With the appearance of cybercrime, computer systems have become an essential source of evidence. However, to investigate a computer and find traces of an attack, the forensic investigator must be familiar with and adopt proper procedures. Already established digital investigation processes define the best tools to use depending on the circumstances of the investigation, data validation/preservation techniques, and stress the value of a chain of custody.

The investigator ensures that no alterations are made to computer systems and their collected data, which must be proven during court procedures. Furthermore, the investigator needs to understand how the machine under investigation stores data to interpret the identified traces. Many vendors use different computer architectures; thus, there are several different ways that data is encoded, such as ASCII and Unicode. The order in which data is stored in memory may differ. This chapter provides a foundation for the digital investigation process and computer systems.

The main objectives of this chapter are as follows:
- To discuss the digital investigation process
- To understand data formats and structures in computer systems
- To learn about numbering systems
- To familiarize yourself with endian ordering, ASCII, and Unicode data formats

3.2 Digital Investigation Process

It was established in a previous chapter (Chapter 1) that the digital forensic investigation process is not rigidly defined, with several researchers

FIGURE 3.1
Digital forensic investigation process.

and organizations proposing their definitions and each shifting the focus of the investigation according to their circumstances. One such definition was given by Carrier et al. [1], as shown in Figure 3.1, and is comprised of three main bidirectional phases: system preservation, evidence searching, and event reconstruction.

3.2.1 System Preservation Phase

During the system preservation phase, actions for data collection and ensuring data preservation are defined. The main goal of data preservation is to ensure, in a provable manner, that evidence integrity is maximized by making a few alterations to the original data as possible. As such, investigators need to ensure that neither their actions nor any processes running on the machine being investigated make alterations to the data. Some important considerations to be made during this first phase of an investigation are the following:

- **Proper live system response**: During an investigation of a live system, data may be altered by processes already running on the machine or by malicious commands sent over the Internet.
- **Order of volatility**: Different subsystems of a computer are characterized by different levels of volatility due to the frequency of updates of the subsystem. The order of volatility for some common computer subsystems is as follows: [CPU Caches, Registers and RAM] > [Virtual Memory] > [Disk Drives] > [Backups and Printouts].
- **Ensuring integrity**: It is vital to establish the integrity of collected data, as not doing so may affect the credibility of a forensic investigation. Alterations to data may occur through many actions, with the alterations being intentional (as a result of malicious actions) or unintentional (simply accessing a file).
- **Establishing and documenting evidence state**: Along with ensuring the integrity of collected data, establishing a chain of custody is of utmost importance.

3.2.2 Evidence Searching Phase

Following data collection and preservation, the next phase of the investigation is tasked with searching for evidence that establishes the occurrence of a security incident (crime). The evidence searching phase defines a cyclic search and hypothesis validation method, where data is scanned to identify evidence that either proves or disproves a hypothesis about the crime committed. For example, the hypothesis may be that a distributed denial of service (DDoS) attack targeted a system. After searching the collected data, traces that indicate a memory overflow for a period of time may support this hypothesis.

Because data may come in many forms, search methods are organized based on data type as follows:

- **Files by file type**: The search target may be a specific PDF file; thus, all files with the ".pdf" extension are investigated.
- **Keyword/string searching**: A file may be identified by using a keyword found in its content.
- **Hash analysis**: Files may be identified by comparing their hash values [2], to entries in databases where the hash values of established software are maintained.
- **Logical/metadata review**: Viewing the metadata of a file to determine when it was last modified.

3.2.3 Evidence Reconstruction Phase

In the final phase of the investigation process, identified evidence from the previous stage is considered and further processed, producing information that can answer important questions that appeared during the investigation. An important distinction that needs to be made in this phase is separating essential and nonessential data. The latter may cause the investigator to lose focus and overlook significant evidence.

3.3 Common Phases of Digital Forensics

Although many digital forensic investigation frameworks have been proposed, as was explained in Chapter 1, it can be observed that four steps commonly found in most frameworks [3] are adequate to describe the investigation process.

- **Acquisition**: In this stage, the investigator takes possession of the computer, either physically or remotely, along with all network mappings and external physical storage devices.
- **Identification**: After collection, this stage covers the actions necessary to identify valuable data and safely extract it or recover it using well-established computer forensic tools and software suits.

- **Analysis and evaluation**: The identified data is then processed and examined to determine its value in a criminal investigation. It is often used to prove misconduct in a corporate setting or a crime in a court of law.

- **Presentation**: Finally, this step involves the presentation of discovered evidence in a law manner understood by lawyers and non-technical staff.

3.4 Numbering Systems and Formats in Computers

A forensic expert needs to interpret the collected data during an investigation before conclusions can be drawn and a hypothesis is proven or disproven. Forensic investigators need to be familiar with the standard numbering systems present in computer systems and used by forensic software [4].

Data in its simplest form exists in binary format, a base 2 number system, with binary files composed of single-valued numbers that can either be "0" (off-state) or "1" (on the state), called bits. The hexadecimal is another numbering system often employed by forensic imaging software to view data on a hard drive. This base 16 number system uses numbers between "0" and "15" with numbers above "9" represented by the first five letters of the English alphabet.

3.4.1 Hexadecimal

In mathematics and computing, hexadecimal is a positional numeral system with a radix, or base, of 16. It uses sixteen distinct symbols, most often the symbols 0–9, to represent values zero to nine, and A, B, C, D, E, F (or A–F) to represent values ten to fifteen. Many computer applications indicate that a number is hexadecimal by adding the "0x" characters in front of the number. One reason for using the hexadecimal representation of a number instead of a decimal is that it is easier to remember them (hex:0x7fffffff, dec:2147483647). A popular hexadecimal editor for the Windows operating system used for data recovery and digital forensics is the WinHex tool [5]. Table 3.1 lists the hexadecimal and decimal versions of numbers 0–16.

Converting numbers between the hexadecimal and decimal systems relies on multiplication and division with base-16. Precisely, to convert hexadecimal into a decimal number, multiply each digit with 16 to the power of its position in the hexadecimal number, starting with 0 at the rightmost digit for the base-10 number 2689, as shown in Figure 3.2. To convert from a decimal to hexadecimal, perform a Euclidean division with the dividend being

TABLE 3.1

Hexadecimal and Decimal
Numbers

Hexadecimal	Decimal
0	0
1	1
2	2
3	3
4	4
5	5
6	6
7	7
8	8
9	9
A	10
B	11
C	12
D	13
E	14
F	15
10	16

$$0xA81 = A*16^2+8*16^1+1*16^0 = A*256+8*16+1*1 = 2560+128+1 = 2689$$

FIGURE 3.2
Converting hexadecimal to decimal.

the decimal number and the divisor the base 16, as shown in Figure 3.3. The arrow in Figure 3.3 indicated the order from most significant (highest power) digit to least significant (the smallest power).

3.4.2 Binary

Binary is a base-2 numeral system that uses "0" to represent zero and "1" to define one. The binary system is used in computing to describe data as a series of zeroes and ones [1,6]. A single binary digit is called a bit and can take the values "0" or "1," and four bits make a nibble, eight bits make a byte, two bytes make a word, two words make a dword (double word), and two dwords make a qword (quadword), as shown in Table 3.2. Historically in computer systems, a single character was encoded by using a byte (8 bits). In many architectures, the smallest addressable amount of memory is the byte [7].

The process of converting numbers between the binary and decimal systems is similar to converting hexadecimal to decimal. The difference

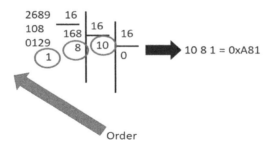

FIGURE 3.3
Converting decimal to hexadecimal.

TABLE 3.2

Binary Data Structures

Number of Bits	Name	Binary
1	Bit	0
4	Nibble	0000
8	Byte	0000 0000
16	Word	0000 0000 0000 0000
32	Dword	0000 0000 0000 0000 0000 0000 0000 0000
64	Qword	0000 0000 0000 0000 0000 0000 0000 0000 0000 0000 0000 0000 0000 0000 0000 0000

is that the number "2" is used for the multiplications and divisions. Both conversions (to and from binary) for the base-10 number "86" are shown in Figures 3.4 and 3.5.

Usually, the smallest allocated space in a computer system is a byte (8 bits), while single bit flags (indicating an option is "on" or "off") are also grouped into bytes. Bytes themselves can be grouped into 2 bytes (word), 4 bytes (dword), and 8 bytes (qword), as shown in Table 3.2. A binary digit can represent two states ("0" and "1"). A binary number is effectively doubled with the addition of a new bit; as such, the possible values that the binary number can represent are also doubled, as demonstrated in Table 3.3. To easily calculate

$$1010110 = 1*2^6+0*2^5+1*2^4+0*2^3+1*2^2+1*2^1+0*2^0+ =$$
$$1*64+0*32+1*16+0*8+1*4+1*2+0*1 = 64+16+4+2 = 86$$

FIGURE 3.4
Converting binary to decimal.

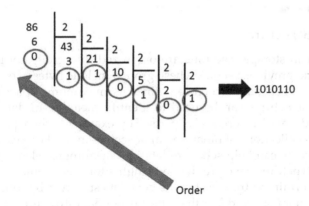

FIGURE 3.5
Converting decimal to binary.

TABLE 3.3

Size of Binary Numbers

Number of Bits	1	2	3	4
Possible states	2	4	8	16
Binary states	0, 1	00, 01, 10, 11	000, 001, 010, 011, 100, 101, 110, 111	0000, 0001, 0010, 0011, 0100, 0101, 0110, 0111, 1000, 1001, 1010, 1011, 1100, 1101, 1110, 1111

the possible states that a binary number can represent, use the n^2 formula, where "n" represents the number of bits.

There are two main strategies for converting hexadecimal to binary, either convert it first to decimal and from decimal to binary (as was shown in the last two sections), or by using the direct approach, convert each hexadecimal digit into its binary equivalent and using a 4-bit representation for each digit (4-bit, because 4 bits can represent 16 numbers 0–15). For example, "0xEF12" is transformed into binary as follows: E(hex)=14(dec)=1110(bin), F(hex)=15(dec)=1111(bin), 1(hex)=1(dec)=0001(bin), 2(hex)=2(dec)=0010(bin). So "0xEF12"="1110 1111 0001 0010."

As an exercise, complete the following conversions:
- 0xDEAB to binary
- 0xDEAB to decimal
- 10111101 to hexadecimal
- 01010101 to decimal

3.5 Data Structures

Data stored in storage mediums are organized into groups of bits, as discussed in the previous section. These groups include bytes, words, dwords, and qwords. Each data structure defines its maximum capacity and data type. Data structures can be used to identify meaningful data entities in storage mediums by forensic experts. For example, when parsing blobs (amorphous collections of binary data) or sectors in hard drives, identifying a data structure can help separate data into meaningful elements. In addition, data structures have pre-defined lengths for their elements, which can assist in determining the offset of the next data structure by considering the starting point (offset) and length of the current data structure.

Thus, the importance of data structures for digital forensic experts is that they allow for the conversion of meaningless data into information and knowledge. Advanced tools used by forensic experts, such as WinHex [5], are capable of detecting and "interpreting" data structures. However, forensic experts need to be able to explain and test a tool's interpretation. This interpretation is the first step in reverse engineering application artefacts [8].

3.5.1 Endianness

Another important information that a forensic expert needs to have about a system they are investigating is endianness [1]. Endianness describes the order in which bytes are stored in memory, with the two main categories being: big-endian and little-endian, as shown in Figure 3.6.

Big-endian ordering defines that the most significant byte is stored in the lowest address in memory. Thus the bytes are read from left to right. On the other hand, little-endian ordering defines that the least significant byte is stored in the lowest address in memory; thus, the bytes are read from right to left. Big-endian ordering has the benefit of being easier to interpret by humans. In contrast, little-endian ordering allows low-level programming optimization, as numbers can be accessed at different data structure lengths.

Originally, the endianness in a computer was determined by the processor architecture. However, in recent years, operating systems have had a

	Decimal	Binary	Hexadecimal
Number	1025	0000 0100 0000 0001	0x0401
Big-endian		00000000 00000000 00000100 00000001	0x00 00 04 01
Little-endian		00000001 00000100 00000000 00000000	0x01 04 00 00

FIGURE 3.6
Example of endianness.

default order, and application developers specify the order for file-level data. Historically, IBM's 370 mainframes, most reduced instruction set computer (RISC)-based computers, and Motorola microprocessors use the big-endian approach. Furthermore, the TCP/IP model in networks also uses the big-endian approach. Thus, big-endian is sometimes called network order. On the other hand, Intel processors (CPUs) and DEC Alphas and at least some programs that run on them are little-endian.

Knowing the endianness of a system is particularly important, as numerical values will be misinterpreted otherwise. For example, date and time in a computer system are calculated as the elapsed time from some pre-determined data (UNIX: seconds since midnight, 1 Jan 2970 UTC, NTFS: number of 100 ns since 1 Jan 1601 UTC, etc.).

3.5.2 Character Encoding

At the lowest level of abstraction, computer data like text, images, and other files are represented by bits. However, at higher levels, standardized numeric codes are used to describe characters. Thus, character encodings that define characters in multiple languages such as ASCII and Unicode emerged [9,10]. The WinHex tool [5] allows for several character encodings through the code page option.

3.5.2.1 ASCII

ASCII, which stands for American Standard Code for Information Interchange, defines 128 characters (from "0" to "127") represented by 7 bits. However, a byte is used (8 bits), which effectively doubles the number of supported characters. Initially, it supported only English characters, but European languages and mathematical symbols were included through extensions.

3.5.2.2 Unicode

On the other hand, Unicode uses a variable bit-length representation, with standard encoding versions being UTF-8, UTF-16, and UTF-32 that use 8, 16, and 32 bits, respectively. Due to its versatile encoding scheme, Unicode supports tens of thousands of characters from multiple languages.

3.6 Data Nature and State

The different states of data describe the location of data and its purpose. There are three primary data states: in transit, in use, and at rest [11]. Data in transit (motion) describes data moved either through a network or between storage mediums in a computer. Data in use represents data that is being accessed by a user or CPU. In this mode, data is temporarily stored in RAM, where it can be accessed faster than other storage media, and it remains there until it is no longer in use or the device is powered off. Finally, data is said to be at rest when it is not in motion or use. In this state, data is stored in non-volatile storage devices such as hard drives and memory cards.

3.6.1 Terms of Data

In this section, two terms about data will be discussed: Metadata and Slack space [11–13].

- **Metadata**: Metadata [12] refers to data about data. A real-world example would be information about a car that is for sale. Information such as the car's colour, location, year of manufacture, type of engine, and transmission type qualifies as metadata. In computer systems, metadata includes network packet headers, time, and date of creation, file index in a hard drive, file size, and creator name.
- **Slack space**: The smallest amount of allocated space in a hard drive that can store data is called a cluster. Clusters often use a default value of 4 KB or can be user-defined. Slack space [13] results from clusters using less than their entire allocated space to store data. In other words, a cluster that has a maximum capacity of 4 KB but stores only 1.5 KB of data is said to have a slack space of 2.5 KB.

3.7 Conclusion

This chapter has discussed the concept of the digital forensic investigation process, expanding from Chapter 1. Furthermore, the hexadecimal and binary numbering systems were introduced, explaining how to convert between them and the decimal. In addition, we explained again ordering and discussed data encodings. Next, the basics of hard disk analysis will be explored.

References

1. Brian Carrier. Digital crime scene investigation process. In: *File System Forensic Analysis*, pp. 13–14. Addison-Wesley Professional, Boston, MA, 2005.
2. Gongzheng Liu, Jingsha He, and Xinggang Xuan. A data preservation method based on blockchain and multidimensional hash for digital forensics. *Complexity*, 2021. doi: 10.1155/2021/5536326.

3. Yunus Yusoff, Roslan Ismail, and Zainuddin Hassan. Common phases of computer forensics investigation models. *International Journal of Computer Science & Information Technology*, 3(3):17–31, 2011.

4. John Sammons. *The Basics of Digital Forensics: The Primer for Getting Started in Digital Forensics*. Elsevier, Amsterdam, 2012.

5. Winhex tool, https://www.x-ways.net/winhex/, 2021.

6. Israel Koren. *Computer Arithmetic Algorithms*. AK Peters/CRC Press, Boca Raton, FL, 2001.

7. Robert W Bemer. A proposal for a generalized card code for 256 characters. *Communications of the ACM*, 2(9):19–23, 1959.

8. Hausi A Müller, Jens H Jahnke, Dennis B Smith, Margaret-Anne Storey, Scott R Tilley, and Kenny Wong. Reverse engineering: A roadmap. In *Proceedings of the Conference on the Future of Software Engineering*, pp. 47–60, Limerick, Ireland, 2000.

9. Jürgen Bettels and F Avery Bishop. Unicode: A universal character code. *Digital Technical Journal*, 5(3):21–31, 1993.

10. Paul Hoffman and François Yergeau. Utf-16, an encoding of iso 10646. Technical report, RFC 2781, February, 2000.

11. Ben Findlay. A forensically-sound methodology for advanced data acquisition from embedded devices at-scene. *Forensic Science International: Reports* 3:100188, 2021.

12. Martin S Olivier. On metadata context in database forensics. *Digital Investigation*, 5(3–4):115–123, 2009.

13. R Vishnu Thampy, K Praveen, and Ashok Kumar Mohan. Data hiding in slack space revisited. *International Journal of Pure and Applied Mathematics*, 118(18):3017–3025, 2018.

4

Fundamentals of Hard Disk Analysis

4.1 Introduction

A cyberattack aims to steal, compromise, or even prevent access to data, which can be proven to be detrimental to users and cause monetary loss. As such, to effectively detect unauthorized changes to data and investigate such events, forensic examiners need to be familiar with the computer subsystems that store and manage data, namely storage media, with an emphasis in this chapter, given on hard disks.

Advances in technology have resulted in more than one type of hard disk, storing data and organizing it differently. This chapter introduces storage media, their structure and functionality, and the booting process that occurs during a computer start-up.

The main objectives of this chapter are as follows:
- To discuss hard disk types
- To understand hard disk settings
- To learn the booting process
- To practise essential Linux commands and python scripts for digital forensics aspects

4.2 Storage Media

According to the technology used for storing data, storage media can be separated into two main categories, rigid platter disk technology, and solid-state technology.

4.2.1 Rigid Platter Disk Technology

Depending on the volatility and technology used, more than one types of storage media exist. This section discusses these storage media and their

FIGURE 4.1
Spinning hard disk components.

technology and techniques to represent, store, and manage data. Magnetic disks are the most commonly employed storage media. Magnetic media is a term that covers many technologies that rely on the electromagnetic interactions between a moving "head," on which an electric charge is applied to produce a magnetic field, and a medium that has a magnetic coating and can become polarised [1]. The part of a magnetic storage media that stores information contains layers of metals or glass, covered oxide-based coating.

This coating becomes magnetized or polarised by applying a magnetic field, created by applying an electric charge to moving "heads," allowing for the recording or accessing information. Small areas that have been magnetized represent a "1" bit, while not magnetized areas represent a "0" bit. Several different types of magnetic storage media have been developed over the years, such as magnetic tapes, floppy disks, and hard drives.

A prominent example of this type of storage is the hard disk drive (HDD) [2]. These storage media are comprised of multiple discs, called platters, which are usually made up of non-magnetic material, such as glass or aluminium. A thin coating of magnetic material is applied on top of the platters, allowing for storing and reading information. A magnetic field is used to the platter, polarising small areas and storing data.

As shown in Figure 4.1, a single HDD incorporates multiple platters, and each platter can be recorded on either side. For that purpose, various sliders and heads are used, with the actuator (axis, arm) providing positional control and a central spindle that connects the multiple platters, providing rotation. Each platter of an HDD is separated into several logical sub-components [3].

- **Tracks**: Concentric circles around the spindle of a platter, each wide enough to be accessed by the read/write head, are called tracks.

- **Cylinder**: Tracks that share a relative position on their respective platters form a cylinder. For example, the collection of outermost tracks on all platters found in an HDD is one cylinder.

- **Sector**: Each platter is separated into wedge-shaped areas called geometric sectors. Each geometric sector is made up of multiple track segments, called sectors.

Sectors are considered the basic unit of storage used to transfer data to and from the HDD and have a fixed size determined by the hardware [3–5]. A sector is a subdivision of a track and stores a fixed amount of user-accessible data [6]. The size of sectors has changed over the years. Traditionally, a sector could store 512 bytes since 1956 and newer hard drives, utilizing what is known as Advanced Format, storing 4096 bytes per sector. In addition to the data stored in a sector, the sector included 50 bytes that stored an error correction code, used to correct errors during the read/write process.

4.2.2 Solid State Technology

A solid-state disk (SSD) is an example of such storage media [7,8]. These storage media rely on semiconductor chips rather than polarising areas of magnetic surfaces, as is done in magnetic media. Solid-state storage is similar to RAM in structure, although it is not volatile and ensures data persistence. Binary values are represented by the flow of electric current through transistors and gates, organized in cells. One cell can store from 1 to 3 bits of information, depending on their design.

SSD uses memory called "flash memory," which is like a RAM and controller. Unlike RAM (volatile), SSD is non-volatile, where data remain if the computer's power is off, as shown in Figure 4.2. SSD uses a grid of electric cells

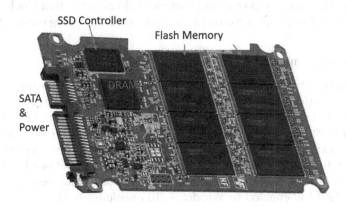

FIGURE 4.2
SSD components.

for reading and writing data. Grids are separated into sections called pages, which are the place where data are stored. These pages clumped together and formed blocks.

The two most essential parts of SSD include the controller and flash memory. First, the controller is an embedded processor that bridges the flash memory components to a host (a computer) and executes the code of the SSD's firmware. The popular functions of the controller include reading, writing, error checking, erasing, garbage collection, encryption, etc. Second, flash memory refers to a particular type of electronically erasable programmable read-only memory. It is a computer memory chip that maintains stored information without requiring a power source. It is often used in portable electronics, such as digital music devices, smartphones and digital cameras, and removable storage devices. The SSD types involve NAND and NOR, non-volatile flash memory technologies, with differences in reading/writing data design.

4.3 Hard Disk Forensic Features

The two important features that affect how hard disks manage data are Garbage Collection and the TRIM command.

4.3.1 Garbage Collection

Garbage collection is a form of automatic memory management that attempts to reclaim "garbage" or memory that was previously occupied by data objects that are no longer in use by programs. In short, SSDs are made up of many blocks, each of which can store bits grouped into pages. If some pages in a block need to be erased because they are no longer valid due to updates, the garbage collection process dictates that the valid pages in that block are copied to a new block, with the data in the previous block being erased [9].

4.3.2 TRIM Command

TRIM is a command issued during file deletion by the operating system to an SSD that informs the latter which data blocks are no longer in use and can be deleted or marked for rewriting. Through TRIM, the SSD is informed about what blocks are invalid and thus, during Garbage Collection, these invalid blocks are not moved to a new block. This has the benefit of providing the SSD with more space while requiring fewer block deletions [9]. The cmd command "fsutil behavior query DisableDeleteNotify" can be used to check whether TRIM is enabled in Windows. To enable (or to disable) TRIM, the previous command is modified as so "fsutil behavior set DisableDeleteNotify 1" ("0" to disable).

4.3.3 Methods of Accessing Hard Disk Addresses

4.3.3.1 Cylinder-Head-Sector (CHS)

Data stored in a sector is called a block of data or simply a block. The CHS was an early addressing scheme for physical data blocks in a storage device [10]. In the CHS scheme, addressing is calculated by CHS, and it starts from 0/0/0, with the size of the hard drive equal to CHS. Early implementations of CHS limited the supported capacity of an HDD to 504 MB, allowing for 1024 cylinders, 16 heads, and 63 sectors [11]. Improvements to the CHS addressing scheme, including an address translation step where the number of cylinders was halved, and the number of heads doubled, allowed for the support of larger storage capacity, reaching 7.85 GB with the maximum number of cylinders 1024 and count of heads 256.

4.3.3.2 Zone-Bit Recording

Zone-bit Recording is an alternative way of creating sectors on an HDD that optimizes the assignment of sectors per track [12]. It assigns tracks into zones based on a track's distance from the disk's centre (spindle). Tracks are split into sectors of equal size, causing outer tracks to have more sectors than inner tracks. Forensic tools often utilize LBA and CHS storage instead of zone-bit, as the previous two methods guarantee that the number of sectors per track is constant on each cylinder.

4.3.3.3 Logical Block Addressing (LBA)

LBA is an alternative addressing scheme to CHS, where data blocks are addressed using an integer index, with the first block being LBA 0, the second LBA 1, and so on, as listed in Table 4.1, derived from a linear translation from the CHS scheme [13]. The majority of HDD produced after 1996 implement LBA data addressing [14]. LBA in 32-bit address mode supports 232 blocks and 2.2 TB of storage, while a 64-bit address mode supports 264 blocks and allows for the theoretical size of 9.44 zettabytes.

4.4 Hard Disk Settings

Settings of an HDD define several characteristics about how the HDD can be accessed by the operating system (OS). Expressly, they represent the disk type, whether it will be split into partitions or volumes, the partition style, and the file system used by the volumes/partitions. Knowing the hard disk settings leads to a better understanding of how the hard drive is managed and allows for the analysis of partitions, volumes, and file systems forensically soundly.

TABLE 4.1

LBA Addressing

LBA	C	H	S
0	0	0	0
1	0	0	1
2	0	0	2
3	0	0	3
4	0	0	4
5	0	0	5
6	0	0	6
7	0	0	7
8	0	0	8
9	0	0	9
10	0	1	0
11	0	1	1
12	0	1	2

The Windows setup program automatically handles these settings for the primary hard disk of the system, although adding other hard drives requires these settings to be selected by the user. In Windows, through the "Disk Manager" [15], the settings mentioned above would be accessed.

4.4.1 Disk Types

Hard disks can be configured in two ways, a basic disk and a dynamic disk. Basic disks [16], in Windows, manage storage drives by using primary and extended partitions. Redundant array of independent disks (RAID) is used, a data storage virtualization technology used to combine physical hard drives into logical units, with the aim of data redundancy, performance, or a combination of the two [17].

Each hard drive supports four primary partitions. Basic disks can be configured in three RAID levels: RAID 0, where data stripping is used; RAID 1, where mirroring is used; and RAID 5, where both stripping and parity are used. Dynamic disk [16] is supported by Windows 2000 and above. It supports 128 partitions per single drive, with these disks divided into volumes.

4.4.2 Partition Architectures

Before a computer can use an HDD, it first needs to be partitioned, a process that determines where data is stored and accessed [18]. Before the partitioning process can occur, a partitioning scheme needs to be selected. The two commonly employed types are Master Boot Record (MBR) and the GUID Partition Table (GPT) schemes. These two partitioning schemes define how the partitioning data is stored in an HDD.

4.4.2.1 MBR and GPT

Chronologically the first of the two, MBR is a partition tabling scheme that is most commonly found in x86 and x64 computers [19]. It supports 4 primary partitions on a single drive, as MBR includes 4 partition entries, or 3 primary and 1 extended partition, with the extended partition further split into 23 partitions. The MBR scheme is compatible with drives of up to 2 TB storage, and it uses hidden sectors that do not belong to any partition to store data that is critical for platform operations [19,20]. Finally, it does not include error check mechanics, such as replication and cyclic redundancy code.

Produced after the MBR, the GPT was developed by Intel in the late 1990s [18], and it is supported by most OSs these days. It supports up to 128 partitions per hard drive and supports volumes up to 18 exabytes. Unlike MBR, GTP stores platform critical data inside its partitions, while it provides superior reliability as it supports replication and cyclic redundancy code for the partition table.

4.4.2.2 Primary and Extended Partitions

A primary partition function as a distinct disk can host an OS. It can be marked as an active partition by using the "0x80" hexadecimal code at the beginning of the partition record [21]. Each primary partition is formatted separately and assigned a unique hard drive letter. MBR supports up to 4 primary partitions, and GPT supports 128 primary partitions.

On the other hand, an extended partition cannot host an OS and must be marked as inactive by using the "0x00" hexadecimal value at the beginning of the partition record [22]. An extended partition is further separated internally into logical drives, with each drive formatted separately.

4.4.2.3 Volumes and Partitions

Volumes and partitions are addressable disk storage sectors that host an OS and its file systems [23]. Partitions are made up of physically consecutive collections of sectors that exist inside a physical hard disk. On the other hand, (logical) volumes may span multiple separate sectors, even spanning different disks, as done when using some RAID versions. However, this separation is not perceived by OSs and file systems. An example of the difference between partitions and volumes can be seen in Figure 4.3.

4.4.3 File Systems

A file system is an underline system applied to storage media that enables the correct storage, management, and later accessing data in the disk [24]. A file system determines where a file is stored, who can access it while maintaining metadata about the file, such as size and name. To set up a file system, a volume or partition needs to be formatted, which causes any pre-existing

FIGURE 4.3
Volumes and partitions.

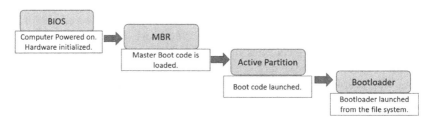

FIGURE 4.4
The boot process.

data to be deleted and creates a new file system. Windows supports several file systems, such as NTFS, FAT32, exFAT, FAT(FAT16), and ReFS [25].

4.4.4 The Boot Process

After powering on a computer, the first software that is executed is called the basic input/output system (BIOS) [26,27], a firmware that is embedded in the hardware and stored in read-only memory. As shown in Figure 4.4, first, the BIOS initializes the computer's hardware and then accesses the MBR to load the master boot code that handles the booting (or bootstrap) process. This, in turn, accesses the MBR partition table to identify and load the active partition that holds the OS code and file system [27].

BIOS provides an interface for OSs to access the hardware through a library of interrupts. During a forensic examination, an investigator removes the suspect's hard drive and clones or images it to ensure that the MBR has not been modified, as changes to the MBR may render the drive unreadable or cause unexpected data loss [19,22].

4.4.4.1 Latest BIOS

The Unified Extensible Firmware Interface (UEFI) is another firmware that handles the bootstrap process and is an alternative to BIOS, with many advantages over the latter [28]. It has a modular design, constituting it extensible and flexible while independent of hardware, platform, OS, and processor. The UEFI does not require a dedicated bootloader (whereas BIOS with MBR does), as the bootloaders are stored in EFI documents and loaded in an EFI shell environment, from where the OS can boot. In essence, the UEFI functions as a software interface between the firmware and the OS and uses an EFI system partition, where the software and EFI files related to the OS bootloader are stored. Primarily, in Windows OS, the UEFI firmware supports GPT.

4.4.4.2 BIOS and MBR

The MBR, located at the first cylinder, head, and record of the hard drive, is the first area that the BIOS accesses to load the necessary code to identify the OS and continue the boot process [29]. The MBR includes the bootstrap code, the partition table, and the disk signature. The bootstrap code is stored in the first 446 Bytes of the MBR and is loaded by the BIOS to identify an active partition that holds an OS.

The partition table holds the starting and ending location (offset) of up to four primary partitions that are either active (bootable) or inactive (not bootable), with their states identified by the first hexadecimal value with the flag "0x80" used for active and "0x00" for inactive. The final two bytes of the disk are called a disk signature and take the value "0x55AA." Two copies of the MBR exist on a disk. The first one is located at sector "0" while the second one is at the last sector. Essential partition elements are given in Table 4.2.

TABLE 4.2

Partition Elements

Byte Offset	Field Length	Field Name and Definition
0x01BE	Byte	**Active partition flag**: Indicates whether the volume is the active partition. **Legal values include**: 00 "Do not use for booting" or 80 "Active partition."
0x01BF	Byte	Start head
0x01C0	6 bits	**Start sector**: Only bits 0–5 are used. The upper two bits, 6 and 7, are used by the starting cylinder field
0x01C1	10 bits	**Start cylinder**: Uses 1 byte and the upper 2 bits from the starting sector field to make up the cylinder value. The starting cylinder is a 10-bit number, with a maximum value of 1023
0x01C2	Byte	**System ID**: Defines the volume type such as NTFS, FAT
0x01C3	Byte	End head
0x01C4	6 bits	**End sector**: Only bits 0–5 are used. The upper two bits, 6 and 7, are used by the ending cylinder field
0x01C5	10 bits	**End cylinder**: Uses 1 byte in addition to the upper 2 bits from the ending sector field to make up the cylinder value. The ending cylinder is a 10-bit number, with a maximum value of 1023.
0x01C6	DWORD	**First sector**: The offset from the beginning of the disk to the beginning of the volume, counting by sectors
0x01CA	DWORD	**Total sectors**: The total number of sectors in the volume

4.5 Essential Linux Commands for Digital Forensics Basics

You practice using Linux commands in Kali using the Kali Linux Virtual machine (the link is provided in Chapter 1). These commands will assist in understanding how to apply digital forensics processes using various tools.

4.5.1 User Privileges

First, before we run any commands, we need to launch the terminal located at the top of the screen. The following commands will be preceded by the prompt root@kali:~#, which you should not type in your commands. To create a new user, we will use the "adduser" command [30]. By default, the Kali OS offers a root account with extended privileges. Create a student user by typing the command:

- root@kali:~# adduser student

Sometimes, you require root privileges to execute some commands/ programs. To do so, your user will need to belong to the sudo group. To add your user to the sudo group, type the following command:

- root@kali:~# adduser student sudo

To run a command in sudo mode, the structure is as follows: "sudo <your command>" [31]. You will then be prompted to add your password. Having multiple users in a machine, you may need at some point to switch between users. To do so, you will use the su (switch user) command [32]. For example, to switch to the user student, the command would be:

- root@kali:~# su student

4.5.2 Linux System

Using the terminal in Linux OS, you may come across commands for which you are uncertain about their options or function. For that purpose, you can use the man command [33] as follows:

- root@kali:~# man adduser

The above command will display the manual page for the adduser command, including its options and arguments. To exit the manual page, just type q.
    While traversing the file system, you may view your current path (full path) by issuing the pwd [34] command without any arguments. To change your directory, use cd [35] (change directory) command while, to view the

contents of your current directory, type ls [36]. In the cd command that follows, we specify that we wish to change the working directory to the one above the current, indicated by the "..". Other directories may be specified, either by using an absolute path (starting from the root directory, like "/root/Desktop") or a relative path starting from the current directory (in the form of "./relative/path/to/target/directory/").

- root@kali:~# pwd
- root@kali:~# cd..
- root@kali:~# ls

To create a new, empty file in the current path, use the touch command [37], while to create a new directory, use the mkdir command [38]. By providing a path to the mkdir command, you can specify the location where your new directory is created. Write permissions may affect whether the command runs successfully or fails. To copy a file from one location of the file system to another, either by using an absolute or relative path, use the cp command [39], which is structured as "cp source destination."

- root@kali:~# touch file_name
- root@kali:~# mkdir new_directory
- root@kali:~# cp /root/my_file my_file1

To remove a file or directory, issue the rm command [40]. In the case of directories, you need to include the -r option, which causes the command to run recursively, removing any subdirectories and files. The echo command [41] is used to echo (print) any text that is added after it as an argument. It can be used to write to a file by redirecting its output by using the ">" character and typing the file name, such as "my_file_overwritten", as listed in the commands below. If you use ">>," instead of overwriting your text to the specified file (as is done with ">"), the text is appended, such as "my_file_appended". To view the contents of a file, use the cat command [42].

- root@kali:~# rm my_file
- root@kali:~# rm -r my_directory
- root@kali:~# echo hello world > my_file_overwritten
- root@kali:~# echo hello world >> my_file_appended
- root@kali:~# cat my_file

Linux files have permissions that limit who can read (r), write (w), and execute (e) them. Permission come in a three-digit code, [(r)(w)(e) (r)(w)(e) (r)(w)(e)) with each element taking either the value "1" (allowed) or "0" (not allowed). Each triplet is considered to be a number in binary form (for instance, "101" allowing read and execute privileges), which is viewed in the file system as a decimal number (the "101" will be represented "5").

The first triplet dictates permissions for the owner of the file, the second for the user group, and the final for the rest of the "world" (every other user). To view the permissions of files under the current directory, use the ls command with the -l option. To change permissions, use the chmod command [43], followed by the permissions number in decimal form and the affected file.

Recall that the permissions are comprised of separate three-digit binary numbers, which are represented in decimal form. As such, to allow read, write, and execute permissions for the owner of a file, you will include the number 700 (111 000 000), while to grant all permissions to everyone, you will use the number 777 (111 111 111).

- root@kali:~# chmod 700 my_file
- root@kali:~# ls -l my_file

When "echo" is not enough, you will need to use a text editor such as nano [44].To launch nano, simply type "nano file_name" to the terminal. To search for text in your opened file, type "Ctr+W," to save your progress, type "Ctr+O" and exit "Ctr+X."

4.5.3 Data Manipulation

To practice some data manipulation techniques, enter the following text into my_file.

1. Derbycon September
2. Shoocon January
3. Bruconer September
4. Blackhat July
5. Bsides *
6. HackerHalted October
7. Hackon April

To scan for the existence of a string in a file, you will use the grep command [45]. For example, running the command "grep September my_file," will return all instances where the string "September" is present in the my_file document. You can combine multiple commands through pipes. Type them one after another to pipe two commands, with the character "|" separating them.

This will cause the output of the first command to be used as input for the second. By piping the previous grep command with the cut command

[46], you can extract specified fields of the returned text, as shown in the command below. The -d command specifies what delimiter to use for tokenizing the text (separate it into fields, -d specifies space as delimiter), while -f indicates which field to return (-f 2 selects the second field/token).

- root@kali:~# grep September my_file
- root@kali:~# grep September my_file | cut -d " " -f 2

Use the sed command to make automatic alterations to a document based on patterns or expressions [47]. In the following command, where we substitute "Blackhat" with "Defcon," the "s" indicates substitution, while the "/" are delimiters. For pattern matching searches, you can use awk [48]. In the current example, the command displayed below will search the first field of the my_file and return instances where it is greater than "5."

- root@kali:~# sed "s/Blackhat/Defcon/" my_file
- root@kali:~# awk "$1> 5" my_file

4.5.4 Managing Packages and Services

In Kali Linux, as in other Debian-based Linux distributions, you have access to the apt (advanced package tool) [49], which allows you to manage software packages. To install a new package, use the apt-get command [50] with the install option, followed by the package in question, as indicated in the command below. To upgrade already installed packages, simply use apt-get with the upgrade option.

- root@kali:~# apt-get install armitage
- root@kali:~# apt-get upgrade

To search whether a program has been installed in your Kali, use the dpkg command [51] with the -l option, followed by the program name.

- root@kali:~# dpkg -l program_name

In Kali OS, multiple services may be running in the background. By using the service command [52], you can start, stop or restart them. You may wish to do so if you make alterations to a service's parameters. To issue the command, you type service followed by the name of the service you wish to affect and then the action you wish to perform, as displayed below.

- "root@kali:~# service apache2 start"
- "root@kali:~# service mysql stop"

4.5.5 Managing Networking

It is not uncommon for a PC to have more than one network interface card, connecting it to one network. To view your active network interface cards and information related to your connections, use the ifconfig command [53]. To view the default gateway through which your host routs traffic to other networks, use route [54]. Should you need to change your network interface card settings, you need to edit the interfaces file, located under the "/etc/network/" directory, as shown in the following command.

- root@kali:~# nano /etc/network/interfaces

Issuing the above command will give you the contents of the interface document, as shown in Figure 4.5. In the above screen, you can see the settings for interface eth0 (the auto keyword precedes interfaces that are to be loaded at boot time). In the "iface eth0 intet static" line, we specify that the interface eth0 will have an IPv4 address, which is statically defined in this document.

To use Dynamic Host Configuration Protocol and get a dynamic address, you would only use the commented line "interface eth0 inet dhcp" and remove the other lines which define IP addresses for the host, the gateway, network, broadcast, and the netmask. After making changes to the interfaces document, you will need to restart the networking service by using the service command.

- root@kali:~# service networking restart

To view your host's network information, including connections, active ports, programs listening on TCP ports, you can use the netstat command

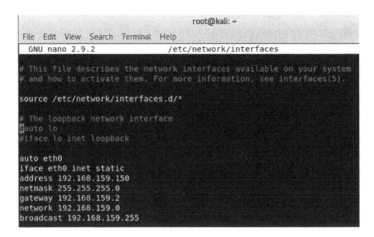

FIGURE 4.5
The file content of network interfaces.

[55]. Useful options of this command include -a that displays both active and non-active TCP sockets, -n shows addresses in numeric format, -t limits the displayed connections to only TCP, and -p shows the process ID and name of the program which listens to each socket.

- root@kali:~# netstat -antp

4.6 Python Scripts for Digital Forensics Basics

Linux systems typically come with interpreters for several scripting languages, such as Python [56] and Perl [57]. You are encouraged to familiarize yourself with the basics of Python 3 that assist you when applying digital forensics techniques. For writing a python script to check the status of some ports for a specific IP address, write the following code, and name the file "mypython.py," shown in Figure 4.6.

The first two lines refer to the python interpreter and library for socket programming, respectively. The following two lines receive user input (IP and port) and store it in variables IP and port. The s variable stores a socket instance that has an (IP address, port number) pair (defined through the AF_INET option) and uses the TCP protocol (determined through the SOCK_STREAM option). Finally, the if condition checks if the port is open for the IP address, port number pare, connecting. To make the file executable by issuing the command "chmod 744 mypython.py." To run the script like so: "python mypython.py," and use IP: "192.168.159.152" and port: "80."

4.6.1 Executing a DoS Attack

First, launch the Windows 10 and Ubuntu Server VMs. In the Ubuntu Server, launch the system monitor from the left bar on the desktop. You should observe close to 0 KB/s traffic in the Network History, as there are no running services.

```
#!/usr/bin/python

import socket
ip=input("IP: ")
port=input("port: ")
s=socket.socket(socket.AF_INET,socket.SOCK_STREAM)
if s.connect_ex((ip,int(port))):
        print ("port/IP are closed")
else:
        print ("port", port, "is open")
```

FIGURE 4.6
Python code to check open and close ports in a network.

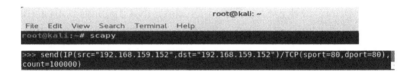

FIGURE 4.7
Scapy code of a distributed denial of service attack in a local area network.

In the Kali VM, you will use Scapy [58] to send fake (forged) requests, effectively performing a volumetric DoS attack against the Windows 10 and Ubuntu Server machines. Scapy is a versatile packet manipulation program written in Python, which can be used to forge or decode packets of a vast number of protocols, transmit them, capture them match requests and replies, and more. To start the program, simply open the terminal and write "scapy." In scapy, type the command depicted in Figure 4.7.

The above code defines a distributed denial of service attack in a local area network. The source and destination IPs and ports are identical, causing the target machine to be stuck in a loop while responding to such a forged packet. The specified IP address is that of the Ubuntu Server VM. The count option indicates the number of packets to be sent, which is "10000." After executing the above command, go back to the Ubuntu Server and view the observed traffic in the Network history.

4.7 Conclusion

This chapter discussed the effects of the hard disk types, settings, and boot tables on the booting process. Furthermore, two main booting firmware, including BIOS and UEFI, were discussed. Essential Linux commands and python scripts were also explained to understand how to apply the digital forensics process. Next, advanced hard disk analysis will be provided for digital forensics perspectives.

References

1. Kazuo Goda and Masaru Kitsuregawa. The history of storage systems. *Proceedings of the IEEE*, 100(Special Centennial Issue):1433–1440, 2012.
2. Martin H Weik. *Magnetic Disk*, pp. 956–956. Springer US, Boston, MA, 2001.
3. Nihad Ahmad Hassan and Rami Hijazi. *Data Hiding Techniques in Windows OS*. Elsevier, Amsterdam, 2017.
4. Marshall Brain. How hard disks work. *Retrieved February*, 13:2008, 2000.

5. Karl Paulsen. *Moving Media Storage Technologies: Applications & Workflows for Video and Media Server Platforms*. Routledge, Oxfordshire, 2012.

6. Andrei Khurshudov. *The Essential Guide to Computer Data Storage: From Floppy to DVD*. Prentice Hall Professional, Hoboken, NJ, 2001.

7. Alan R Olson, Denis J Langlois, et al. Solid state drives data reliability and lifetime. *Imation White Paper*, pp. 1–27, 2008.

8. Michael Cornwell. Anatomy of a solid-state drive. *Communications of the ACM*, 55(12):59–63, 2012.

9. Kent Smith. Garbage collection. SandForce, Flash Memory Summit, Santa Clara, CA, pp. 1–9, 2011.

10. Xiaodong Lin. *Introductory Computer Forensics: A Hands-on Practical Approach*. Springer International Publishing, Cham, 2018.

11. Brandon J Hopkins and Kevin A Riggle. An economical method for securely disintegrating solid-state drives using blenders. *Journal of Digital Forensics, Security and Law* 16(2):1, 2021.

12. Computer forensics: Investigating file and operating systems, wireless networks, and storage (CHFI). Computer Hacking Forensic Investigator. Cengage Learning, 2016.

13. Wasim Ahmad Bhat. Capacity barriers in hard disks: Problems, solutions and lessons. *International Journal of Information Technology*, 12:155–163. 2020. doi: 10.1007/s41870-018-0144-x.

14. Jongmin Gim, Youjip Won, Jaehyeok Chang, Junseok Shim, and Youngseon Park. Dig: Rapid characterization of modern hard disk drive and its performance implication. In *2008 Fifth IEEE International Workshop on Storage Network Architecture and Parallel I/Os*, pp. 74–83. IEEE, Baltimore, United States, 2008.

15. Overview of Disk Management, https://docs.microsoft.com/en-us/windows-server/storage/disk-management/overview-of-disk-management, 2021.

16. Bill Ferguson. *MCDST: Microsoft Certified Desktop Support Technician Study Guide: Exams 70–271 and 70–272*. Wiley, Hoboken, NJ, 2006.

17. Adam Leventhal. Triple-parity raid and beyond. *Communications of the ACM*, 53(1):58–63, 2010.

18. Bruce J Nikkel. Forensic analysis of gpt disks and guid partition tables. *Digital Investigation*, 6(1–2):39–47, 2009.

19. Cory Altheide and Harlan Carvey. Chapter 3: Disk and file system analysis. In Cory Altheide and Harlan Carvey (editors), *Digital Forensics with Open Source Tools*, pp. 39–67. Syngress, Boston, MA, 2011.

20. Pawan K Bhardwaj, Chapter 6: Managing hard disks with the diskpart utility. *How to Cheat at Windows System Administration Using Command Line Scripts, How to Cheat*, pp. 171–201. Syngress, Burlington, 2006.

21. Pawan K Bhardwaj, Chapter 5: Maintaining hard disks. *How to Cheat at Windows System Administration Using Command Line Scripts, How to Cheat*, pp. 135–170. Syngress, Burlington, 2006. https://www.elsevier.com/books/how-to-cheat-at-windows-system-administration-using-command-line-scripts/bhardwaj/978-1-59749-105-1

22. David Day. Chapter 7 - Seizing, imaging, and analyzing digital evidence: Step-by-step guidelines. In: Babak Akhgar, Andrew Staniforth, and Francesca Bosco (editors), *Cyber Crime and Cyber Terrorism Investigator's Handbook*, pp. 71–89. Syngress, Burlington, 2014.

23. Michael Hasenstein. The logical volume manager (lvm). White paper, 2001.

24. Tammy Noergaard. Chapter 5: File systems. In: Tammy Noergaard (editor), *Demystifying Embedded Systems Middleware*, pp. 191–253. Newnes, Burlington, 2010.

25. Johnson M Hart. *Windows System Programming*. AddisonWesley Microsoft Technology Series. Pearson Education, London, 2010.

26. Jeff Tyson. How bios works. How Stuff Works.viitattu 19.10. 2012. Saatavissa: http://computer. howstuffworks. com/bios.htm, 2008.

27. Amar Rajendra Mudiraj. Windows, linux and mac operating system booting process: A comparative study. *International Journal of Research in Computer and Communication Technology*, 2(11):2278–5841, 2013.

28. Vincent Zimmer, Michael Rothman, and Suresh Marisetty. *Beyond BIOS: Developing with the Unified Extensible Firmware Interface* (Third Edition). De Gruyter, Incorporated, Berlin, 2017.

29. Littlejohn Shinder and Michael Cross. Chapter 4: Understanding the technology. In: Littlejohn Shinder and Michael Cross (editors), *Scene of the Cybercrime* (Second Edition), pp. 121–200. Syngress, Burlington, 2008.

30. adduser, https://linuxize.com/post/how-to-create-users-in-linux-using-the-useradd-command/, 2021.

31. sudo, https://www.linuxfromscratch.org/blfs/view/9.1/postlfs/sudo.html, 2021.

32. su, https://lowfatlinux.com/linux-switch-user-su.html, 2021.

33. man, https://www.geeksforgeeks.org/man-command-in-linux-with-examples/, 2021.

34. pwd, https://factorpad.com/tech/linux-essentials/pwd-command.html, 2021.

35. cd, https://www.guyhowto.com/linux-cd-command/, 2021.

36. ls, https://www.linuxteck.com/basic-ls-command-in-linux-with-examples/, 2021.

37. touch, https://www.howtoforge.com/tutorial/linux-touch-command/, 2021.

38. mkdir, https://www.geeksforgeeks.org/mkdir-command-in-linux-with-examples/, 2021.

39. cp, https://www.geeksforgeeks.org/cp-command-linux-examples/, 2021.

40. rm, https://www.linuxhelp.com/rm-command, 2021.

41. echo, https://www.tecmint.com/echo-command-in-linux/, 2021.

42. Cat, https://www.geeksforgeeks.org/cat-command-in-linux-with-examples/, 2021.

43. chmod, https://www.howtogeek.com/437958/how-to-use-the-chmod-command-on-linux/, 2021.

44. nano, https://phoenixnap.com/kb/use-nano-text-editor-commands-linux, 2021.

45. grep, https://www.thegeekstuff.com/2009/03/15-practical-unix-grep-command-examples/, 2021.

46. Cut, https://linuxize.com/post/linux-cut-command/, 2021.

47. sed, https://linuxhint.com/50_sed_command_examples/, 2021.

48. awk, https://www.digitalocean.com/community/tutorials/how-to-use-the-awk-language-to-manipulate-text-in-linux, 2021.

49. apt, https://www.tecmint.com/best-open-source-linux-text-editors/, 2021.

50. apt-get, https://linuxize.com/post/how-to-use-nano-text-editor/, 2021.

51. dpkg, https://www.debian.org/doc/manuals/debian-tutorial/ch-editor.html, 2021.

52. service, https://docs.rackspace.com/support/how-to/command-line-text-editors-in-linux/, 2021.

53. ifconfig, https://www.tecmint.com/ifconfig-command-examples/, 2021.

54. Route, https://www.thegeekstuff.com/2012/04/route-examples/, 2021.

55. netstat, https://www.tecmint.com/20-netstat-commands-for-linux-network-management/, 2021.
56. Python, https://www.python.org/, 2021.
57. Perl, https://www.perl.org/about.html, 2021.
58. Scapy, https://scapy.net/, 2021.

5

Advanced Hard Disk Analysis

5.1 Introduction

In Chapter 4, we discussed the essential hard disk components. We also introduced the Master Boot Record (MBR) located at the first sector of a disk that stores the location of the partitions and how to access them during the booting process. However, the MBR is not the only type of partitioning scheme, with GUID Partition Table (GPT) being the popular alternative. A forensic expert needs to understand how these partition schemes are formed and how they work.

This information is often crucial during the recovery of deleted partitions that may hide important information. In this chapter, we will delve into the differences between discs using the MBR and GPT schemes. Furthermore, we will describe the components of the GPT and how it points to partitions in a hard disk. Finally, we will discuss the process of recovering deleted partitions of hard disks that employ MBR or GPT partitioning.

The main objectives of this chapter are as follows:

- To discuss the differences between MBR and GPT disks
- To understand the GPT table partition and its header
- To learn how to recover deleted partitions from MBR or GPT disks

5.2 Hard Disk Forensic Concepts

During a typical digital forensic investigation, entire hard disks are extracted from suspect computers and utilized to locate useful data in the form of traces and evidence. Thus, it is of vital importance to determine the structure of partitions and volumes so that files can be extracted effectively. This is done through partition tables, where the start, end, and type (file

TABLE 5.1

Example of a Partition Table

Start	End	Type
0	99	FAT
100	599	NTFS

system type) are stored for the OS to utilize, as can be seen in Table 5.1. Attacking the partition tables and altering entries of legitimate partitions is one way that hackers attempt to hide or otherwise make data unavailable to investigators [9,12,].

Such tampering with the partition table may be located through consistency checks [6]. To perform such checks, the starting and ending locations of all known partitions are first gathered. If there are unused sectors between the partitions or the last partition and the end of the disk, this indicates that data may be hidden there, and the sectors should be analysed. If, on the other hand, the partitions overlap, with one partition starting before another has ended, or one partition encompasses another, then this implies that the partition table has been corrupted. In general, unpartitioned space between legitimate partitions, gap, and reserved partitions should be investigated for valuable data.

A partition can be recovered by assuming that each identified partition had a file system, and then searching for known patterns or "magic" values that indicate the type of partition [6,7] and can be found at specific locations of a partition, sometimes in its first sector. For example, in a FAT file system, the "0x55AA" value is stored in byte 510 of sector 0 (a logical partition address), and in Ext2, the "magic" value is "0XEF51." By identifying partition type information and looking for the main elements of the MBR and GPT schemes present in a hard drive, including allowed lengths, it becomes possible to rebuild such partitions.

5.3 DOS-Based Partitions

Being synonymous with MBR, DOS partitions are commonly found in many Windows and Linux installations, with Mac computers being the exception, as they often employ Apple Partition Map or GPT [16]. The MBR partition table [11] starts at sector 0, with the offset of 446 bytes that store valuable code for the boot process, and is 64 bytes long with 4 entries 16 bytes long each. Each partition entry stores (1) the partition type, (2) the active flag, (3) the start and end of partition, and (4) partition length. However, the maximum partition size depends on the file system.

TABLE 5.2

Structure of MBR Partition Table Entry

Offset	Size	Description
00	1 byte	Status indicator (active/bootable)
01	3 bytes	Starting CHS (head: 8 bytes, sector: 6 bits, cylinder: 10 bits)
04	1 byte	System indicator (file system type)
05	3 bytes	Ending CHS (same as start)
08	4 bytes (dword)	Starting LBA address
0C	4 bytes (dword)	Number of sectors (in partition)

The structure of a single 16-bit long MBR partition table entry is given in Table 5.2. The first byte indicates the status of the partition entry, with a value of "0x80" indicating an active partition and "0x00" an inactive (not bootable). The values for starting/ending cylinder, head, and sector, and the number of bits used, are limited to the ranges 0–1023, 0–255, and 0–63, respectively. Finally, the starting logical block address (LBA) address indicates the partition's first sector's beginning and is used instead of the CHS (cylinder, head, and sector).

5.3.1 Revisited MBR

Figure 5.1 depicts the MBR partition table from Windows 10 virtual machine. The first rectangle is the first partition table entry, and the second one is the second partition table entry. The first byte is "00" for the first partition, indicating an inactive (or not bootable) partition. The next three bytes indicate the start CHS (10, 8, and 6 bits, respectively) address for partition 1. However, the values are in little-endian format, as the partition entry originated from a Windows OS. The next byte indicates the type of file system, in this case, a dell maintenance partition, followed by three bytes for the end CHS address ("0x03 3F FE" in big-endian).

The initial LBA is given by the next dword indicating a start at sector 63 ("0x00 00 00 3F"), and the total number of sectors by the remaining dword (4 bytes). Starting at sector 63, the initial LBA address would indicate that sectors 1–62 (as sector 0 is used by MBR) would contain additional boot code or hidden data. The values are derived similarly for partition 2, with partition 1 and 2 values given in Table 5.3 in a hexadecimal format.

```
00 00 00 00 00 2C 44 63  9E 6E C9 9D 00 00 00 01   ......,DclnÉl....
01 00 DE FE 3F 03 3F 00  00 00 C5 FA 00 00 80 00   ..Þþ?.?...Åú..I.
01 04 07 FE FF FF 04 FB  00 00 BB 42 FB 06 00 00   ...þÿÿ.û..»Bû...
00 00 00 00 00 00 00 00  00 00 00 00 00 00 00 00   ................
00 00 00 00 00 00 00 00  00 00 00 00 00 00 55 AA   ..............U ª
```

FIGURE 5.1

Instances of partition consistency check.

TABLE 5.3

Values of Windows 10 VM Partition Table Entries

Offset	Size	Partition 1	Partition 2
00	1 byte	00 (inactive)	80 (active)
01	3 bytes	00 01 01	04 01 00
04	1 byte	DE	07
05	3 bytes	03 3F FE	FF FF FE
08	4 bytes (dword)	00 00 00 3F	00 00 FB 04
0C	4 bytes (dword)	00 00 FA C5	06 FB 42 BB

5.4 GPT Disks

A disk that uses the GPT partition scheme starts with a protective MBR sector to ensure backwards compatibility with systems that require MBR [13]. In structure, the GPT is similar to the MBR. Their main difference is that GPT includes a backup partition structure and the primary partition structure for redundancy and security. The primary structure is located at the beginning of the disk, right after the protected MBR block, and the backup partition structure (entries followed by GPT header) is at the end of the disk. GPT uses LBA to identify these structures, with the protective MBR located at LBA 0.

The GPT header is located at LBA 1, followed by LBA 2–33, where the partition entries 1–128 can be found. When the OS boots through BIOS, it accesses the first sector to load the bootload code stored in the MBR. However, OSs that use the GPT do not access the MBR through BIOS but instead utilize the UEFI [13]. One of the 128 allowed primary partitions under the GPT scheme needs to be an EFI system partition to boot correctly. All required files for the booting process are stored files, including the bootloaders for all installed OSs.

In Table 5.4, the GPT partition header values are listed, including length in bytes. The GPT header includes cyclic redundancy codes (CRCs) for both the header content and the partition entries. This ensures that any unexpected modifications to either the header or the partition entries can be detected. The LBA of the partition entries is usually set to 2, as the MBR is in LBA 0 and the GPT header in LBA 1. The GPT header includes several null bytes (set to "0"), used for padding because the sector size exceeds the bytes of the header (92 bytes of header values, with, e.g. 512 bytes for a sector, thus 420 null values). Figure 5.2 depicts the header values of the GPT present in the Windows 10 VM using the Active Disk Editor tool.

A disk using GPT includes 128 partition entries [5,14], and each partition entry is 128 bytes long (defined in GPT header). The partition array, made up of the GPT partition entries, starts right after the header, and the components

TABLE 5.4

GPT Partition Header

Length	Data Structure
8 bytes	Signature ("EFI PART," 0x45 46 49 20 50 41 52 54)
4 bytes	GPT revision number
4 bytes	Head size (bytes)
4 bytes	Header CRC (from offset 0 to header size)
4 bytes	Reserved (value=0)
8 bytes	Current LBA address (usually 1)
8 bytes	Backup LBA address
8 bytes	First usable LBA (where the first partition can start)
8 bytes	Last usable LBA (where the last partition must end)
16 bytes	Disk GUID
8 bytes	Starting LBA of entries (where partition array entries start, usually 2)
4 bytes	Number of partition entries in the array
4 bytes	Entry size (size of single partition entry, usually 128 bytes)
4 bytes	CRC of partition array
*	Reserved (null padded)

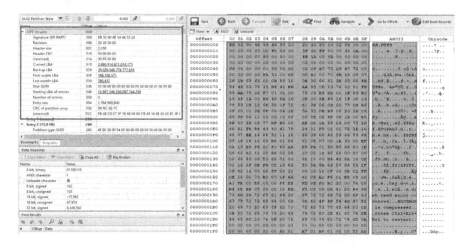

FIGURE 5.2
Windows 10 VM GPT header values using the active disk editor tool.

of a GPT partition entry are given in Table 5.5. The first two fields are 16 bytes long each and define a partition type GUID and a GUID for the partition, respectively, in little-endian form.

The following two fields denote the start and end addresses in LBA format (8 bytes each), followed by attribute flags (8 bytes) and the partition name (72 bytes). The attribute flags include flags for EFI, legacy BIOS and accessing status (read-only, auto-mount, hidden). In Figure 5.3, the first GPT partition

TABLE 5.5

GPT Partition Entry

Offset	Length	Data Structure
0	16 bytes	Partition type GUID
16	16 bytes	Unique partition GUID
32	8 bytes	First LBA address (little-endian)
40	8 bytes	Last LBA address (little-endian)
48	8 bytes	Attribute flags
56	72 bytes	Partition name

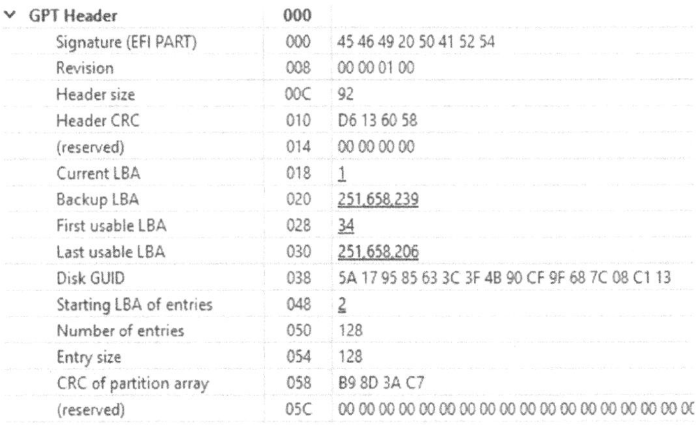

∨ GPT Header	000	
Signature (EFI PART)	000	45 46 49 20 50 41 52 54
Revision	008	00 00 01 00
Header size	00C	92
Header CRC	010	D6 13 60 58
(reserved)	014	00 00 00 00
Current LBA	018	1
Backup LBA	020	251,658,239
First usable LBA	028	34
Last usable LBA	030	251,658,206
Disk GUID	038	5A 17 95 85 63 3C 3F 4B 90 CF 9F 68 7C 08 C1 13
Starting LBA of entries	048	2
Number of entries	050	128
Entry size	054	128
CRC of partition array	058	B9 8D 3A C7
(reserved)	05C	00 00 00 00 00 00 00 00 00 00 00 00 00 00 00 00 0C

FIGURE 5.3
Windows 10 VM GPT header values.

entry, which is the EFI System Partition of the Windows 10 VM, plays a vital role in the booting process. Any unused partitions will have entries with a "0" value.

5.5 Forensic Implications

Reformatting a hard disk, or in other words, converting [15,17] from an MBR partition scheme to a GPT and vice versa in Windows, can be done through the command line using the diskpart tool [1] or the Disk Management [2] software. However, a pre-requisite to convert between MBR and GPT is that

```
00000000336 10 EB F2 F4 EB FD 2B C9   E4 64 EB 00 24 02 E0 F8   ëòòëÿ+Éädë $ àø
00000000352 24 02 C3 49 6E 76 61 6C   69 64 20 70 61 72 74 69   $ ÃInvalid parti
00000000368 74 69 6F 6E 20 74 61 62   6C 65 00 45 72 72 6F 72   tion table Error
00000000384 20 6C 6F 61 64 69 6E 67   20 6F 70 65 72 61 74 69    loading operati
00000000400 6E 67 20 73 79 73 74 65   6D 00 4D 69 73 73 69 6E   ng system Missin
00000000416 67 20 6F 70 65 72 61 74   69 6E 67 20 73 79 73 74   g operating syst
00000000432 65 6D 00 00 00 63 7B 9A   97 D8 5C 8F 00 00 80 00   em  c{││Ø\  █
00000000448 01 04 07 0F 60 93 00 08   00 00 00 20 03 00 00 00    `█
00000000464 41 94 07 0F E0 FF 00 28   03 00 00 38 6D 74 00 00   A█ àÿ (    8mt
00000000480 00 00 00 00 00 00 00 00   00 00 00 00 00 00 00 00
00000000496 00 00 00 00 00 00 00 00   00 00 00 00 00 0█ 55 AA            █Uª
```

FIGURE 5.4
MBR partition table example.

the disk should not contain any pre-existing partitions/volumes, thus converting comes with data loss and may be used to hinder an investigator's attempts to recover data. Furthermore, administrator rights are required to complete these tasks.

To link partitions to the system, OSs utilize and leave forensic traces of disk/partition GUIDs. Useful information may be determined from these traces, such as date/time stamp and access time. Furthermore, in GPT disks, hiding data between partitions may be easier than in MBR due to the more significant number of partition entries (128).

5.6 Practical Exercises for Computer Foundations (Windows)

In this section, the computer foundations of windows for digital forensics are discussed. More importantly, the components of a hard drive using the WinHex tool in Windows 10 are explained.

An example of an MBR partition table, with the entries given in Figure 5.4, is presented. The analysis is given in Table 5.6 in the form of questions and answers.

5.6.1 WinHex Tool

After launching the Windows 10 VM, navigate to the WinHex folder found on the desktop and run it with administrator privileges by right-clicking the ".exe" file and selecting "Run as administrator." WinHex was developed by X-Ways Software Technology AG of Germany [3,4]. WinHex is a powerful tool that functions as an enhanced hex editor. It can be used for data analysis, editing, and wiping, and it can also be used for evidence gathering as a forensic tool. WinHex can:

TABLE 5.6

MBR Partition Table Analysis

Questions	Answers
How many bytes are in a DOS-based partition table?	64
How long is each partition table entry?	16
How many can primary partitions be added?	4
What is the starting offset of the partition table?	446
What are the starting offsets of each partition entry?	1st: 446 2nd: 462 3rd: 478 4th: 494
How many partitions are in Figure 5.4?	Two, because the first two 16-byte entries are not-null
Which partition is the active partition?	The first, because it starts with "0x80" (active)
Assume that, this partition table is stored on a hard disk that includes 255 cylinders and 1025 sectors. What is the total size of the hard disk?	The total size of a hard drive is calculated as follows: #_Cylinders*#_Heads*#_Sectors*Sector_Size= 255*2*1025*512=261375 MB
What is the LBA starting sector of the first partition?	"0x00 00 08 00," 2048 in decimal
What is the size (in bytes) of the first partition?	The number of sectors: "0x00 03 20 00," 204800 in decimal. Size: sectors*size_of_sector = 204800*512 = 104,857,600 bytes
Is there any unpartitioned space in-between the two partitions? How do you know?	No. The first partition ends at (LBA+length)=(2048+204800)= 206848, second partition starts at "0x00 03 28 00," 206848
What happens to the partition table when you delete a partition?	The partition table entry (16 Bytes long) is set to null ("0")
What happens to the partition itself (i.e. the partition's data) when you delete a partition entry?	Nothing

- Parse and edit hard drives (FAT, NTFS), CD-ROMs, DVDs, RAM, and other storage media
- Interpret 20 data types
- Join, split, analyse, and compare files
- Search, clone, replace, and image drives

- Recover data and encrypt files (128-bit strength)
- Generate hashes and checksums
- Edit partition tables, boot sectors, and other data structures using templates, and gather free and slack space

In WinHex, navigate to the "Tools" option and select "Start Centre." In the new window that appears, options for opening files, disks, memory, and directories are available. Select each option to see what is displayed. WinHex, like most hex editors, provides three main panels, an offset (in hexadecimal) that indicates the position in memory where each column of data starts and ends. A 16-byte hexadecimal display of the data that is being inspected and a code page that uses some encoding to display the data in a human-readable fashion (for example, ANSI ASCII).

In WinHex, navigate to the "File" tab, "Open," and select "E:\Photos\ IMG1870," as shown in Figure 5.5. By clicking on the offset column, the display alternates between hexadecimal and decimal format. In the hexadecimal display, the data of the selected image are presented, in hexadecimal format, with each row consisting of 16 bytes and each byte represented as two digits in the hexadecimal system, as 1 byte is equal to 2 nibbles (4 bits) and a nibble can represent a single hexadecimal digit.

A byte whose decimal value is "65" is displayed as "41" in hexadecimal notation $(4 \cdot 16+1=65)$ and refers to the letter "A" in text mode. The ASCII character set defines the capital letter A to have the decimal value of 65. In WinHex, navigate to "File" and select "new" to create a new file. In the

FIGURE 5.5
Example of opening images in WinHex tool.

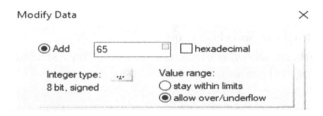

FIGURE 5.6
Modify byte in a new file in WinHex.

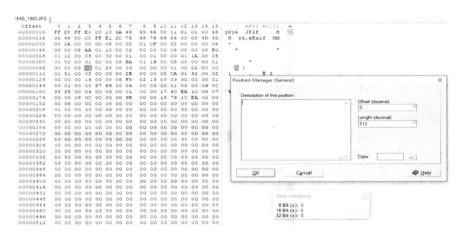

FIGURE 5.7
Modify byte in the original image.

"Create New File" that appears, set the "Size" equal to "1" byte. Right-click the newly created file, select "Edit" and then "Modify data," then add the number "65," as shown in Figure 5.6.

Extended sections of the photo may be selected for editing by right-clicking on any byte of the hexadecimal display and selecting "Edit Position." For example, Figure 5.7 illustrates this process in the previously selected "IMG-1870" photo, with offset set to "0" and length set to "512," with colour set to "Yellow."

Some hexadecimal numbers are reserved as control codes, as shown in Table 5.7. The code page is the text representation of the data using the ASCII character set as used in non-Unicode Windows applications. In contrast, MS-DOS and Windows use the IBM ASCII character set. A different character set can be selected by left-clicking on the "ANSI ASCII" at the top of the code page. As an example, to see what changes in the display, choose UTF-8.

Another useful panel in WinHex is the status bar. The status bar displays vital statistics about a file, such as a filename, creation date, and time. Furthermore, the status bar also shows the file's state (original/modified)

TABLE 5.7

Reserved Hexadecimal Code

Hex	Control Code	Hex	Control Code
00	Null	10	Data link escape
01	Start of header	11	Device control 1
02	Start of text	12	Device control 2
03	End of text	13	Device control 3
04	End of transmission	14	Device control 4
05	Enquiry	15	Negate acknowledge
06	Acknowledge	16	Synchronous idle
07	Bell	17	End of transmission block
08	Backspace	18	Cancel
09	Horizontal tab	19	End of medium
0A	Line feed	1A	Substitute
0B	Vertical tab	1B	Escape
0C	Form feed	1C	File separator
0D	Carriage return	1D	Group separator
0E	Shift out	1E	Record separator
0F	Shift in	1F	Unit separator

and undo levels. As an exercise, do the following: right-click on the status bar, select "Default Edit Mode" and set it to "Editable" in the hexadecimal display, select the first three bytes then "Edit," "Modify data" and add "11" (hexadecimal number) then "ok." Save the file as "image_edit.JPG" and then open the edited image (full path: "E:\IMG_edit.JPG").

The Data interpreter window translates hexadecimal values at the insertion point into decimal based on chosen data types. Default data types include 8, 16, and 32-bit signed, while additional options are available by either double-clicking on the "Data Interpreter" window or selecting "Options" and then "Data Interpreter." Different data types include viewing Assembly language code, data formats, double, GUID, and other integer types.

5.6.2 Recovering Deleted Partitions

The two main partition schemes that we have studied are the MBR and GPT. Each has unique characteristics that can be used to identify it in a hard disk. The MBR partition table consists of four primary partitions or three primary and one extended partition, with the extended partition allowing for multiple logical partitions. Each logical partition includes an extended boot record (EBR) similar in structure to the MBR, with their differences being that for the EBR, the first 446 bytes in the MBR that hold a boot code are usually nulled (set to "0"). Only the first two out of the four records are used. In both cases (MBR and EBR), the final two bytes of the partition table function as a signature ("0x55 AA") and can be used to detect partitions both primary and logical during a forensic investigation.

The GPT partition table, on the other hand, defines multiple partition table entries, usually up to 128 partitions, with four entries in each sector (sector: 512 Bytes). Each GPT partition table entry defines the starting and ending LBA of the blocks, allowing for the recovery of lost partitions. If the information that can locate a partition is modified or cleaned in a partition table, Windows will be unable to find it, causing the partition to be considered missing. Recovering the partition under these circumstances requires utilizing partition recovery tools, such as WinHex and MiniTool Partition Wizard.

Partitions may also be detected and recovered manually by considering the MBR/GPT partition table values and determining the sector size and starting/ending of LBA. For example, Figure 5.8 depicts the partitions available in the Windows 10 VM hard disk, with the top showing the partitions before and the bottom after partition 8 is deleted. It should be noted here that the hard disk has partition gaps between the partition, unpartitioned space, EFI and Microsoft reserved partitions, which usually are between 1 and 200 MB and do not reach or exceed GB storage. If there is a large number of sectors missing, then probably either data files or partition(s) may be either deleted or unmounted.

In Figure 5.8, notice the first box where partition 8 is still present and the red box where it has been deleted. Specifically, notice the difference in sectors in the two instances shown in the purple boxes. The second (bottom) box

Name	Ext.	Size	Created	Modified	Record changed	Attr.	1st sector ▲
Start sectors		1.0 MB					0
Partition 1	FAT32	200 MB					2,048
Partition 2	?	128 MB					411,648
Partition 3 (C:)	NTFS	80.2 GB					673,792
Partition gap		730 KB					168,780,364
Partition 4	NTFS	464 MB					168,781,824
Partition gap		1.0 MB					169,732,096
Partition 5 (F:)	NTFS	19.5 GB					169,734,144
Partition gap		1.0 MB					210,692,096
Partition 6 (E:)	NTFS	9.8 GB					210,694,144
Partition 7 (I:)	FAT32	4.9 GB					231,174,144
Partition gap		1.0 MB					241,408,000
Partition 8	NTFS	4.9 GB					241,410,048
Unpartitioned space		2.0 MB					251,654,144

Partitioning style: GPT							
Name	Ext.	Size	Created	Modified	Record changed	Attr.	1st sector ▲
Start sectors		1.0 MB					0
Partition 1	FAT32	200 MB	•	Partition 8 was deleted, Observe the			2,048
Partition 2	?	128 MB		differences between the number of			411,648
Partition 3 (C:)	NTFS	80.2 GB		sectors of unpartitioned space,			673,792
Partition gap		730 KB		partition gap, and Partition 7,			168,780,364
Partition 4	NTFS	464 MB		compared with the total number of			168,781,824
Partition gap		1.0 MB		sectors. Many sectors are missed.			169,732,096
Partition 5 (F:)	NTFS	19.5 GB					169,734,144
Partition gap		1.0 MB					210,692,096
Partition 6 (E:)	NTFS	9.8 GB					210,694,144
Partition 7 (I:)	FAT32	4.9 GB					231,174,144
Partition gap		1.0 MB					241,408,000
Unpartitioned space		2.0 MB					251,654,144

FIGURE 5.8
Example of detecting deleted partition.

displays a notable difference in starting LBA addresses between the partition gap and the unpartitioned space. This difference, multiplied by the sector size, gives us the space (in bytes) between the two entities and indicates the presence of a removed partition.

5.6.3 Investigating Cyber Threat and Discovering Evidence

In a previous example, an image was modified by altering the first three bytes, resulting in a corrupted file. In real-world crimes, hacking techniques that may target files to hinder their integrity include Null Byte Injection [18] and Data Poisoning attacks [10]. By reviewing the original file and comparing it with the modified file, the differences between the two files can be detected in the hexadecimal display and the code page of WinHex. Another well-established method for checking a file's integrity is by using hash values. If the two files produce different hash values, then the files are not identical.

To perform this check-in WinHex, navigate to the "Tools" tab, then "Compute Hash," and select "MD5 (128 bit)." MD5 is a hash function that receives variable input and produces a 128-bit hash value [8]. Additionally, the files' metadata can be reviewed, with differences in the "Creation time" and "Last write time" indicating that a file has been modified, as shown in Figure 5.9.

By using the "Compare Data" feature of WinHex, found under the "Tools" tab, a report may be produced, indicating what parts of the files were modified. The created report shows all changes in hexadecimal format, with the first column corresponding to the first data source and the second column to the second file, as in Figure 5.10. As an exercise, modify the edited photo using the original hexadecimal values and save it as a new file to see if the image can be viewed.

5.6.4 Hard Disk Analysis

Investigating hard disks is another essential task handled by forensic experts. In WinHex, navigate to "Tools" and select "Open Disk." From the "Logical

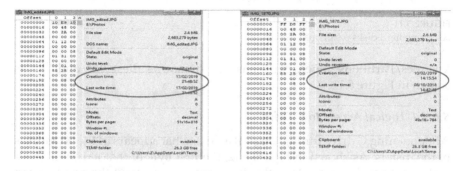

FIGURE 5.9
Modify byte of an image in a new file in WinHex.

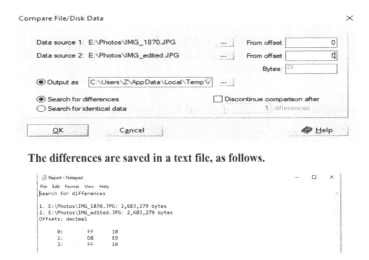

The differences are saved in a text file, as follows.

FIGURE 5.10
WinHex file comparison and report.

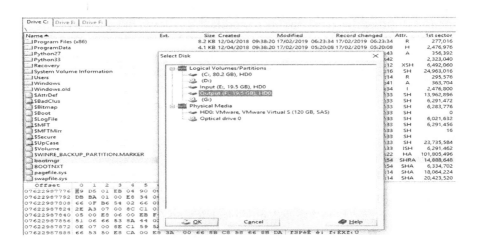

FIGURE 5.11
Accessing hard disks partitions.

Volumes/Partitions" open partitions "C" (windows files), "E" and "F" as shown in Figure 5.11.

5.6.4.1 Logical Access to C Drive

After opening C drive through logical access, what appears as the 00 address of the disk is. As shown in Figure 5.12, the first byte of that partition's boot sector and not the first byte of the disk. Information about the partition can be found in the status bar.

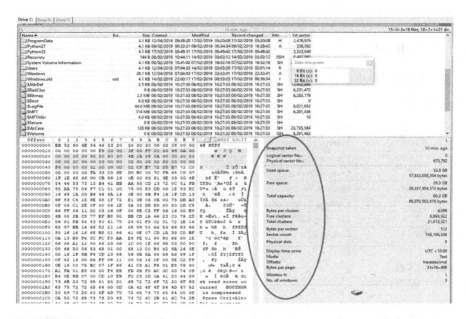

FIGURE 5.12
Logical access to C drive using WinHex.

5.6.4.2 Accessing Drive as Physical Media

Navigate to the "Tools" tab and then to "Open Disk," "Physical Media" and select "HD0: VMware, VMware virtual S (120 GB, SAS)." In the window that appears, click on "start sectors." The data that starts from offset 00, with value "33," is the physical start of the hard drive, starting at head 0, cylinder 0, and sector 0. That location stores the bootstrap code, partition table, and entries for the disk.

To see the template of partitions in Physical Media (HD0), right-click on any partition and select "template." A display appears, as shown in Figure 5.13, indicating various MBR offsets and values. As seen in the figure, the MBR ending sector is "7F."

The MBR consists of the first 512 bytes of information (512 bytes per sector) of the hard disk in sector 0. The first 446 bytes of information contain the bootstrap code. The following 64 bytes are the partition table – four 16-byte records – , and the last two bytes include a signature that identifies the end of all boot sectors: "55 AA."

5.7 Conclusion

This chapter has discussed the GPT partitioning scheme, header and partition table structures, and differences with the MBR. Furthermore, it

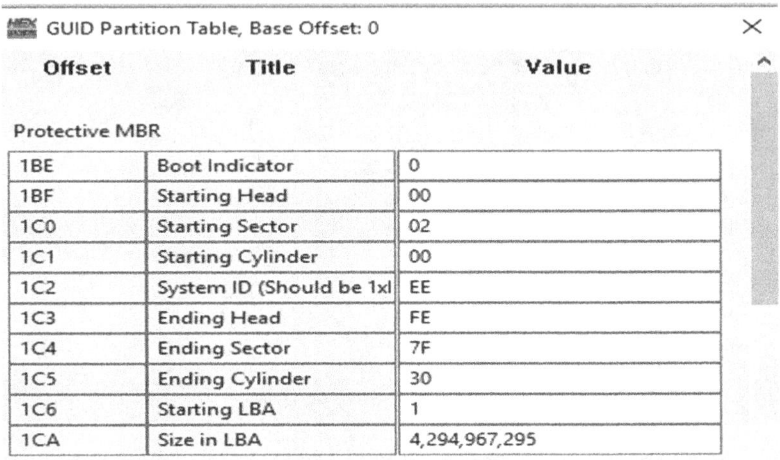

Offset	Title	Value
1BE	Boot Indicator	0
1BF	Starting Head	00
1C0	Starting Sector	02
1C1	Starting Cylinder	00
1C2	System ID (Should be 1x	EE
1C3	Ending Head	FE
1C4	Ending Sector	7F
1C5	Ending Cylinder	30
1C6	Starting LBA	1
1CA	Size in LBA	4,294,967,295

GUID Partition Table, Base Offset: 0

Protective MBR

FIGURE 5.13
Viewing templates.

discussed how to interpret the characteristics of the two partition schemes in a digital forensic setting to recover deleted partitions. Next, the file system in Windows OS will be investigated.

References

1. Diskpart tool, https://docs.microsoft.com/en-us/windows-server/administration/windows-commands/diskpart, 2021.
2. Micorsoft Disk manager, https://docs.microsoft.com/en-us/windows-server/storage/disk-management/overview-of-disk-management, 2021.
3. Winhex tool, https://www.x-ways.net/winhex/, 2021.
4. X-ways, http://www.x-ways.net/forensics/, 2021.
5. André Årnes, ed. *Digital Forensics*. John Wiley & Sons, Hoboken, NJ, 2017.
6. Brian Carrier. Digital crime scene investigation process. In: *File System Forensic Analysis*, pp. 13–14. Addison-Wesley Professional, Boston, MA, 2005.
7. Pabitra P Choudhury. *Operating Systems: Principles and Design*. Prentice-Hall of India Pvt. Limited, Delhi, 2009.
8. Hans Dobbertin. Cryptanalysis of md5 compress. *Rump Session of Eurocrypt*, 96:71–82, 1996.
9. Nihad A Hassan and Rami Hijazi. *Data Hiding Techniques in Windows OS: A Practical Approach to Investigation and Defense*. Elsevier Science, Amsterdam, 2016.
10. Ruoxi Jia, Ioannis C Konstantakopoulos, Bo Li, and Costas Spanos. Poisoning attacks on data-driven utility learning in games. In *2018 Annual American Control Conference (ACC)*, pp. 5774–5780. IEEE, Milwaukee, WI, USA, 2018.

11. Xiaodong Lin. *Introductory Computer Forensics: A Hands-on Practical Approach.* Springer International Publishing, New York, 2018.

12. Kevin Mandia and Chris Prosise. *Incident Response & Computer Forensics* (2nd Edition). Security Series. Mcgrawhill, New York, 2003.

13. Scott Mueller. *Upgrading and Repairing PCs: Upgrading and Repairing_c21.* Pearson Education, London, 2013.

14. Bruce J Nikkel. Forensic analysis of gpt disks and guid partition tables. *Digital Investigation*, 6(1–2):39–47, 2009.

15. Michael Palmer. *MCITP Guide to Microsoft Windows Server 2008, Server Administration, Exam #70–646.* MCTS Series. Cengage Learning, Boston, MA, 2010.

16. Roderick W Smith. *LPIC-2 Linux Professional Institute Certification Study Guide: Exams 201 and 202.* Wiley, Hoboken, NJ, 2011.

17. Neil Smyth. *Windows Server 2008 R2 Essentials.* Payload Media, Cary, NC, 2011.

18. Natasa Suteva, Aleksandra Mileva, and Mario Loleski. Finding forensic evidence for several web attacks. *International Journal of Internet Technology and Secured Transactions*, 6(1):64–78, 2015.

6

File System Analysis (Windows)

6.1 Introduction

In the previous chapters, I have investigated the hard disk and partitioning schemes, their characteristics, and structures. In this chapter, we will be describing the unused space between partitions in hard disks. Furthermore, I will be discussing file system categories and learning how to analyse file contents.

The main objectives of this chapter are as follows:

- To discuss the file system, its categories, contents, metadata, file names, and applications
- To understand slack space, free space, and inter-partition space
- To learn how to analyse and wipe file contents
- To practice hashing for data integrity, data acquisition using the FTK imager tool, and data analysis using the Autopsy tool

6.2 What Is a File System?

A file system functions as a roadmap, a level of abstraction, between the operating system (OS) and the data stored in the storage media [25]. It informs the OS where and how to store and find data, where and how to retrieve stored data and data about the data (metadata), while it provides hierarchical data storage through files and directories. As file systems are closely connected to specific OSs, their implementations may differ; for example, two OSs may delete a FAT file differently [18]. The internal structure of a file, however, is application-dependent. Several file systems have been developed over the years, with some examples being:

- FAT
  - FAT12, FAT16, FAT32

DOI: 10.1201/9781003278962-6

- exFAT (extended FAT)
- NTFS
- HFS+(Mac)
- EXT2, EXT3, EXT4 (Linux, Unix, etc.)
- High-Performance File System (HPFS – IBM's OS/2)
- ZFS (Oracle, originally Sun microsystems)
- ISO9660 (CD-ROMS)
- Distributed file systems

Knowledge of file systems allows a forensic investigator to access data in storage media and reconstruct events. By accessing the file systems in the Windows 10 VM through WinHex [3], you will see that each file system has a template with particular elements, as seen for NTFS and FAT in Figure 6.1. In a FAT partition, the OS utilizes a boot sector to boot and load a partition, with a boot sector copy maintained and used if the original boot sector is corrupted. Furthermore, there are usually two copies of the table that maintains information used when storing/retrieving files in the partition, named FAT1 and FAT2, with FAT2 being a copy of FAT1 for redundancy [2].

In addition, FAT makes use of a root directory [12], where information about all files and directories is stored, such as file name, address of the first cluster of the file, and size of the file. To retrieve a file, the file system first accesses the root directory to locate the file's first cluster and then utilizes the FAT1 table for the remaining clusters.

In an NTFS partition, similar to the FAT, a boot sector at the beginning of the partition allows OS to load the file system and its data, with the

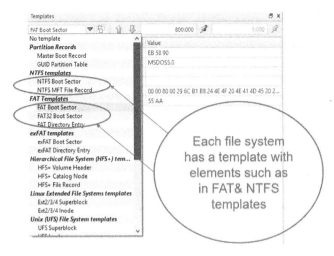

FIGURE 6.1
Entry elements of file systems using active disk tool.

boot sector (sometimes called the volume boot sector), mirrored at the end of the partition, for redundancy and security [5,9]. In NTFS, everything is considered a file, and all relevant information about all documents and directories, such as file name, clusters in HD where it resides, permissions, and other metadata, are stored in the master file table ($MFT). The $MFT is mirrored if the original $MFT is corrupted, and both copies are stored in the boot sector.

6.2.1 File System Reference Model

To be able to compare different file systems and their mechanics, a reference model should be used. This chapter will use such a reference model provided by Carrier et al. [10]. Carrier's file system reference model defines five categories, file system, content, metadata, file name, and application. The file system category provides general information about the file system or a map of the file system. The content category provides information about the files and how they are stored in clusters or blocks in a hard drive. The metadata category describes information about the files, like date of creation, size, and security permissions. The file name category provides data that function as a human interface to the files, metadata, and content. The application category provides a special collection of features about the files that are not essential to storing and accessing files.

Information in the file system category affects the file name, metadata, and content categories, with the file name category interacting with the metadata category and the metadata category in tern interacting with the content category. It stores essential information related to storing and organizing data, like file system data structures that make up the files stored in a hard drive and maps where vital information about the file system is stored, such as the $MFT for NTFS and root directory entries for FAT. Other information stored in the file system category includes the file system version, the OS that created it, creation date and time, and file system label.

The content category, typically the largest part of the file system, stores data related to file and directory content. Files are stored in equal-sized "chunks" of space, comprised of groups of sectors, with each file system using a different name for these "chunks" such as allocation units, clusters, blocks, etc. Tracking of the allocated and unallocated data units that make up the file contents is handled by the file system category.

Usually, a file system does not provide addresses for individual sectors; instead, it groups several sectors into logical data units that have different names in different OSs [17], like "clusters" in Windows/DOS, "blocks" or "allocation units" in UNIX/Linux and "allocation blocks" in MAC OS. The size of these data units is often specified during the file system formatting process, and file systems have default values for the units that can be used, with typical sizes for the data units being 1024 bytes, 2048 bytes, and 4096 bytes [6].

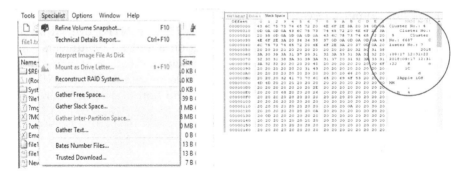

FIGURE 6.2
Example of using WinHex to examine slack space in partitions.

6.2.2 Slack Space

The number of bytes used for addressing limits the number of data units that can be defined and used in a disk. Furthermore, a single data unit can hold the contents of a single file (broken down to chunks though it may be). This means that data units at the end of the volume will include sectors that do not store any information and remain unallocated. This is called "slack space" [13,16]. For example, assume that a FAT partition size is 16,384 bytes and the data unit size is 4096 bytes. If that partition stores a file with a size of 14,256 bytes, the slack space is 16,384 − 14,256 = 2128 bytes, and the partition holds 16,384/4096 = 4 data units. An example of using WinHex to examine slack space is shown in Figure 6.2.

Slack space can be categorized into "Volume slack" and "File System slack" [10,26]. Volume slack refers to the unused space (sectors) between the end of a file system and the end of the partition, where that file system exists. File system slack, on the other hand, refers to the unused space at the end of a file system that is not allocated to a cluster, with an example being when a file is stored in multiple clusters, the remaining unused space of the final cluster that stores the last part of the file. Data can be hidden in both kinds of slack, with volume slack able to store virtually unlimited data, as it can be resized. The file system slack depends on the cluster size, where the maximum slack space in a cluster can be estimated by counting the sectors in a cluster − 1.

6.2.3 Free and Inter-Partition Space

Two other instances where space in a volume is not active use are "free space" and "inter-partition space." OS keeps track of what areas (data units) are in active use through pointers in a file allocation table. When a file is deleted, its pointer is unassigned and not the actual space [24]. Thus, the data unit maintains any data it previously stored. However, it is marked as available

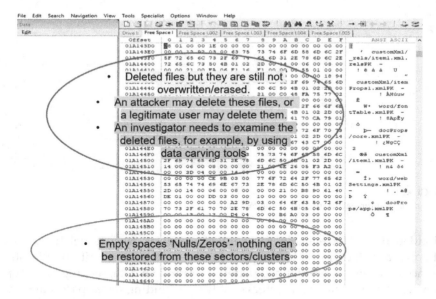

FIGURE 6.3
Example of free space using WinHex.

space by the OS and thus can be assigned to a new file, overwriting any pre-existing data in the process, as depicted in Figure 6.3.

On the other hand, inter-partition space occurs when the hard disk is separated into a partition (see Figure 6.4), and not all sectors get allocated to a partition [19]. This results in some sectors being unallocated and the space being unreachable by the end-users, making it a prime choice for hiding data. Investigators need to check and analyse both spaces, as they may hold useful hidden data.

6.2.4 Content Analysis

Depending on the target and the outcome, analysis of data in the content category can take many forms [10]. In "Data unit viewing," an investigator analyses and recovers files and their metadata by investigating specific data units, accessing them through an already known LBA in the volume. When investigators know what kind of data, they should retrieve but do not know where they are located, they perform a "Logical file system-level searching" by using keywords and strings and searching each data unit. On the other hand, when it is known that the data being sought is stored in unallocated space, investigators access the "data unit allocation status" and seek unallocated units and their content, utilizing forensic tools to extract them.

It was previously mentioned that deleting a file does not immediately erase its data from the data units that stored it previously, thus allowing an investigator to search seemingly unused sectors for useful, potentially

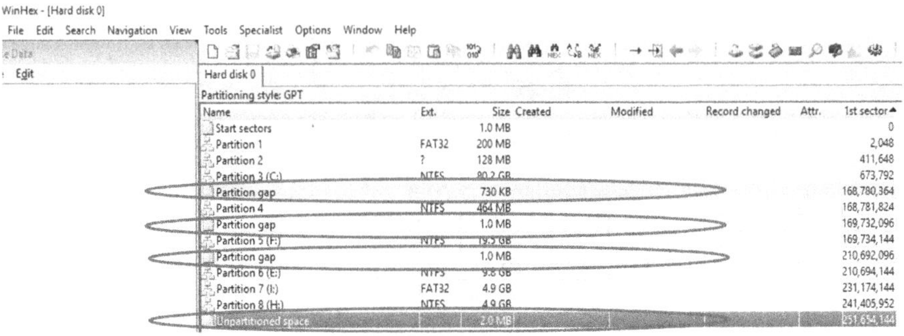

FIGURE 6.4
Example of inter-partition space using WinHex.

hidden data. However, in the content category, "secure deletion," also known as data wiping techniques, can be used to overwrite unallocated space with zeroes, pre-defined patterns, or random data, seeking to destroy any electronic data on all sectors of a drive, including slack space, eliminating sensitive documents and thus hindering an investigation [10, 22,23]. In addition, data may be hidden in "bad sectors" by adding otherwise functional sectors with hidden data to the "bad sectors" list. File systems often track bad sectors and use "Slipping" or "Remapping" to either change the LBA mapping and skip the defective sectors or relocate the defective sector to a spare one.

6.3 Methods for Recovering Data from Deleted Files

6.3.1 Data Carving and Gathering Text

Through specialized tools, such as WinHex [3] and techniques, it is possible to recover data from deleted files. A process used in digital forensics to extract data from a disk drive without utilizing the assistance of the file system that created the files that are being retrieved is called "Data/File carving" [15]. Through data carving, files and fragments may be recovered from unallocated data points without using any file information, which can prove helpful when directory entries or MFT are corrupted, missing, or otherwise unusable. Through WinHex, data carving is possible by navigating to "Tools," "Disk Tools" and selecting "File Recovery by Type."

To extract the ASCII/Unicode characters from hard drives, the "Gather Text" utility can be used. This utility harvests text such as emails, documents, lists of Web pages that have been viewed and downloaded to the PC, and remaining text in slack space and free space that was once part of a file.

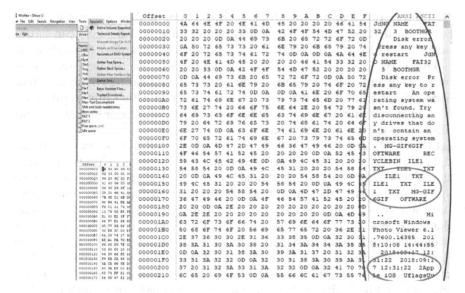

FIGURE 6.5
Example of gathering text using WinHex.

To extract text from disk drives with WinHex, navigate "Select Tools" and select "Gather Text," as depicted in Figure 6.5.

6.3.2 Metadata Category Analysis

The Metadata Category stores data that describes the data, such as date/time of creation, storage location, and access privileges [10]. Metadata is often stored in some kind of table, either of fixed size (FAT) or dynamic length (MFT of NTFS). File deletion is marked in the metadata category. In other words, when a file is deleted, its content often remains intact, with the data units marked as unallocated in the metadata category and some metadata changing, such as in FAT where the first letter of the file name is changed to "0xE5." Deleting files does not change NTFS metadata.

Performing analysis on the metadata category has some essential characteristics that need to be addressed. Compressing and encrypting files make accessing their contents difficult and their metadata are typically left intact [7]. Furthermore, logical file search becomes possible, using information stored in metadata entries, allowing for the search of files in a logical order, rather than the order in which the files are stored in the volume [10]. In addition, when trying to identify the contents of a deleted file, an investigator should search for metadata entries that are no longer allocated to a file, as file-name table entries can be reused before the corresponding metadata entries are overwritten, resulting in metadata for deleted files to be still present in the disk and often overlooked.

Different types of analysis become possible by utilizing metadata attributes. For instance, through data and time stamps, investigators can figure out when a file was created/accessed or otherwise modified through what is known as timeline analysis [11]. Moreover, permission-based analysis is enabled through security/permissions metadata indicating which user had access rights to the files that are being investigated. Other important metadata include a file's data unit addressing, data unit size information, and allocation status of its metadata entries.

Consistency checks may also be useful when used in tandem with metadata analysis. For example, checking that the allocated data units that store the content of files are enough to hold their data and that there is no excessive space (data units) allocated to them [10]. Additionally, each data unit should have at most one metadata entry pointing to it. If there is more than one metadata entry pointing to the same data unit, this is an indication of an error. It should also be verified that particular types of files created by file systems (such as sockets) do not have data unit allocations. Finally, wiping utilities may target metadata of files, setting their values to zeroes or shifting entries in an attempt to hinder or thwart file recovery actions by investigators [20].

6.3.3 File Name and Application Category Analysis

Users can use files' names to access a file's content and metadata in the file name category. Analysis in the file name category is tightly related to metadata analysis, with the former often preceding and even prompting the latter [10]. Common types of analysis in the file name category include searching files by name, including deleted files, since often file systems do not clear the name of a file after deleting it, and by extension. Wiping utilities that operate in the file name category by either zeroing out the file name and metadata pointer entries of a file or shuffling the names of files.

Some file system data exist in the application category and are not essential to the file system itself. They are special features for file system "applications," with application-based file systems maintaining and tracking the file system data. An example of an application-based file system is a "Journaling File System" [14]. The task of a journaling file system is to record changes that are about to be made to the file system (metadata) in a data structure called a "journal" that has not yet happened. Thus, in the event of a crash, if there are changes that were halted, the journal can be used to continue the tasks and complete the changes.

6.4 Practices for Using Hashing and Data Acquisition

This section explains the data acquisition process using the FTK imager tool in Windows 10. Furthermore, analysing the produced images using the Autopsy tool is also described.

6.4.1 Prerequisite Steps for Doing the Following Practical Exercises

Do the following:

1. Launch the Windows 10 VM.
2. Format Partition (H:) and Partition (I)
3. Right-click on Partition (H:) and select Format
4. Right-click on partition (I:) and select Format
5. Open Partition (E:) Copy the folders' general files' and "Photos" and Paste them to Partition (H:) and Partition (I:)
6. Delete some files from Partition (H:) and Partition (I:)

6.4.2 Data Acquisition

During the data acquisition phase of a forensic investigation, forensic experts recover potentially useful sources of evidence by extracting a forensic image from collected media, such as hard drives, removable hard drives, USB, CDs/DVDs, and other storage media obtained from servers, PCs, laptops, gaming consoles and other devices [8]. In the later phases of the investigation, this acquired image is used for extracting and analysing evidence rather than the original source, with the forensic image verified against the original to ensure that the former is an exact duplicate of the latter.

A disk image is a file that replicates the contents, structure, and metadata found in a storage medium, effectively being an exact duplicate of the original [19]. Although it might sound similar to a normal backup, a forensic disk image maintains the integrity of the exact storage structures of the original, which is of utmost importance as it ensures the integrity of a forensic investigation. Suppose it cannot be asserted that a disk image is an exact duplicate (content and structure) of the original data source. In that case, the integrity of any identified evidence is in jeopardy and could be inadmissible in a court of law [21]. As such, it is acquiring a forensic disk image file from a target device is the first step of any digital forensic investigation.

6.4.2.1 The FTK Imager Tool

The Forensic Toolkit (FTK) tool [4] is imaging software created by AccessData, capable of creating local and remote images. The free version enables sonly local imaging and can be used to acquire images of storage devices such as hard disks, USB, CD drives, and individual files. To start the FTK imager, in the Windows 10 VM, navigate to the "Access Data FTK Imager" in the Desktop. A screen will appear, similar to Figure 6.6.

To create a disk image, simply navigate to "File" and select "Create Disk Image." Options allow the capture from memory and imaging of individual items. Next, you are prompted to select the Source Evidence Type, with

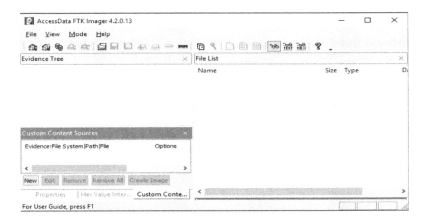

FIGURE 6.6
The FTK imaging tool.

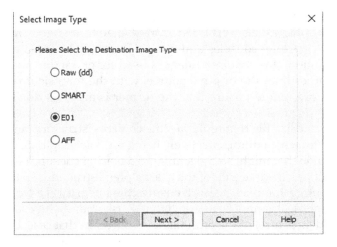

FIGURE 6.7
Image types using FTK imager.

"Logical Drive" indicating that a portion of the hard disk will be imaged, while a "Physical Drive" produces the image of the entire hard disk. Then, a window with a list of available drives will be provided, allowing for the selection of the volume to be imaged. In this case, select "H:\Train_NTFS" and then click "Finish."

A "Create Image" window will appear, displaying the image source (volume to be imaged). Click the "Add" button to select the image type and the destination of the image file that will be generated. As shown in Figure 6.7, the image type refers to the kind of image that will be produced, with four possible choices, Raw (dd) that refers to "disk dump" that can be used in Windows and Linux, SMART that stores the metadata in a separate text file for easier viewing, E01 (our choice for this lab) that stores hash values, case description information and other details in the same file (used by EnCase an enterprise digital forensics program) and AFF that stores all data and metadata in a single file.

After selecting the type of image and output, a window will prompt you to add information about the case for which this image will be produced. This option is a great help during an investigation, as keeping track of evidence and having detailed notes is extremely important. To do so, enter the following: Case Number: "1," Evidence Number: "1," Unique Description: "Imaging H Drive on my local hard disk," Examiner: your name. The folder where the imaging file will be stored is "F:\My-Images" and give the file a suitable name, like "H Drive-Local Disk." The "Image Fragment Size" allows you to specify the size (in megabytes) of the fragments in which the image file will be split, as shown in Figure 6.8. This option is ideal when working with large disk images. If the value given is greater than the size of the hard disk, then only one file is created with the original hard disk's size.

The next option specifies compression of the image, although some image types (like "dd") cannot be compressed, while others (like "E01") can. After finishing with this window, make sure that the "Verify images after they are created" is selected, as it allows for the automatic creation of hash values, which will later be used to verify the image's validity. After the imaging process is completed, an "Image Verify Results" window will appear, displaying both MD5 and SHA1 hash values of the image, as depicted in Figure 6.9. Storing these hash values is essential, as they should be reproducible after the analysis to assert the integrity of the data. Selecting the

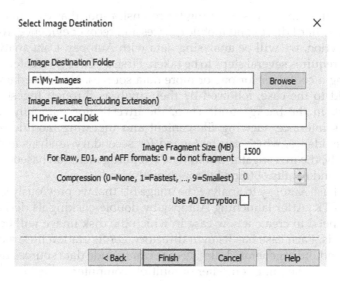

FIGURE 6.8
Options of image destination using FTK imager.

FIGURE 6.9
Image verify results using FTK imager.

"Image Summary" option displays a detailed summary of the generated image, including the produced hash values. In addition, this information has been copied to a text document located in the same directory as the disk image.

6.4.2.2 Hard Disk Analysis Using the Autopsy Tool

Autopsy [1] is an open-source digital forensics tool that can analyse data sources like hard drives and mobile devices and recover evidence from them. In this section, we will be analysing data with Autopsy. Data analysis with Autopsy requires several steps to be taken. First, a "Case" is created, with the case being a container for one or more data sources. Next, the data sources are added to the case, followed by their analysis through Ingest modules that work in the background. Then, the investigator manually navigates Autopsy's interface, viewing file content and ingesting module results to identify evidence, which can be tagged for secondary analysis and reporting. Finally, the investigator can initiate a report generation based on tagged evidence and results obtained through analysis.

We will use Autopsy to analyse the image file that we previously generated through FTK. After launching Autopsy by double-clicking its desktop icon, you first need to create a new case in which the disk image will be opened for analysis. Each case has its own directory, configuration files, a database, reports, and other generated files. A case can include data sources from multiple drives of the same computer or multiple computers.

To start, select "Create New Case," and in the "New Case Information" window that appears, enter the case name "case1" and base directory "C:\data." After clicking next, you will need to add more information like the case number and investigator's name (enter "1" and your name, respectively). After entering the requested information (and clicking "Finis"), a minute is required to create the database for the case and all the necessary modules. Next, select "Disk Image or VM file as Data Source," as shown in Figure 6.10, to load the E01 file that FTK generated.

You will then be prompted to enter the path to the data source. The disk image was stored in the "F" drive, with its full path being "F:\data-acquisition of E drive\img1.E01." The time-zone should be "GMT +10:00 Australia/Sydney," as that value was used when the image was created. However, the time-zone information does not affect the metadata of the image and can be set to "GMT+0:00" if the investigator is uncertain. The next step is to ingest module selection. Ingest modules perform analysis on the selected data source, parting they contend, calculating hash values, and searching for keywords. Navigate to "Select All" and then "Next," as in Figure 6.11.

After clicking "Finish," a new window will appear, displaying the data source, file types, and deleted files, including their metadata (in "Views") and ingest module results like email messages and EXIF files, as shown in Figure 6.12.

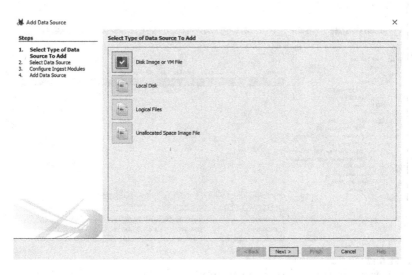

FIGURE 6.10
Selecting data source in Autopsy.

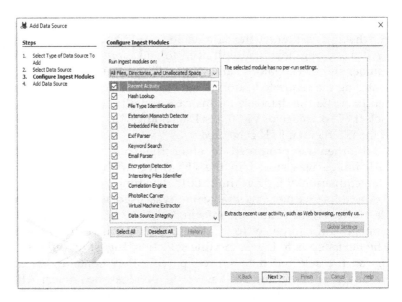

FIGURE 6.11
Selecting ingest modules in Autopsy.

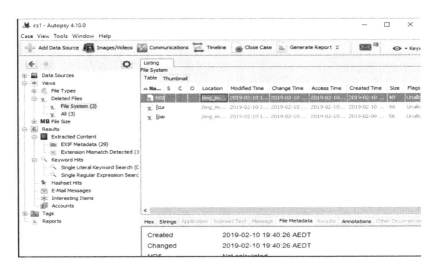

FIGURE 6.12
Analysis and results of a data source in Autopsy.

6.5 Conclusion

This chapter has discussed several file system categories and their components. Furthermore, the concepts of slack space, free space, and inter-partition space were also explained. Several types of analysis and file content wipe practices were presented. Various methods for recovering data from deleted files were also discussed. Practical exercises for hashing, data acquisition using the FTK image tool, and hard disk analysis using the Autopsy tool were also covered.

References

1. Autopsy tool, https://www.autopsy.com/, 2021.
2. Dave Kleiman, Kevin Cardwell, Timothy Clinton, Michael Cross, Michael Gregg, Jesse Varsalone, and Craig Wright. Chapter 14: Forensics investigation using encase. *The Official CHFI Study Guide (Exam 312–49)*, pp. 617–673. Syngress, Rockland, 2007. https://www.sciencedirect.com/science/article/pii/B9781597491976500158
3. Winhex tool, https://www.x-ways.net/winhex/, 2021.
4. Ftk imager tool, https://accessdata.com/product-download/ftk-imager-version-4-5, 2021.
5. Scott Anson, Steve Bunting, Ryan Johnson, and Steve Pearson. *Mastering Windows Network Forensics and Investigation*. ITPro collection. Wiley, Hoboken, NJ, 2012.
6. Jim Aspinwall. *PC Hacks: 100 Industrial-Strength Tips & Tools*. Hacks Series. O'Reilly Media, Newton, NJ, 2005.
7. Diane Barrett, Martin M Weiss, and Kirk Hausman. *CompTIA Security+ SYO-401 Exam Cram: Comp Secu SY04 Auth ePub _4*. Exam Cram. Pearson Education, London, 2015.
8. Ahmed Bouridane. *Imaging for Forensics and Security: From Theory to Practice*. Signals and Communication Technology. Springer US, New York, 2009.
9. Steve Bunting and William Wei. *EnCase Computer Forensics: The Official EnCE: EnCase?Certified Examiner Study Guide*. Wiley, Hoboken, NJ, 2006.
10. Brian Carrier. Digital crime scene investigation process. In: *File System Forensic Analysis*, pp. 13–14. Addison-Wesley Professional, Boston, MA, 2005.
11. Harlan Carvey. Chapter 7: Timeline analysis. In: Harlan Carvey (editor), *Windows Forensic Analysis Toolkit* (Fourth Edition), pp. 211–251. Syngress, Boston, MA, 2014.
12. Vlado Damjanovski. Chapter 9: Video management systems. In: Vlado Damjanovski (editor) *CCTV* (Third Edition), pp. 322–359. Butterworth-Heinemann, Boston, MA, 2014.
13. Larry E Daniel and Lars E Daniel. Chapter 29 deleted data. In: *Digital Forensics for Legal Professionals*, pp. 195–205. Syngress, Boston, MA, 2012.

14. Justin Davies, Roger Whittaker, and William von Hagen. *SUSE Linux 9 Bible.* Wiley, Hoboken, NJ, 2005.
15. Joe Fichera and Steven Bolt. *Network Intrusion Analysis: Methodologies, Tools, and Techniques for Incident Analysis and Response.* Elsevier Science, Amsterdam, 2012.
16. Nihad Ahmad Hassan and Rami Hijazi. Chapter 4: Data hiding under windowsÂ R os file structure. In: Nihad Ahmad Hassan and Rami Hijazi (editors), *Data Hiding Techniques in Windows OS,* pp. 97–132. Syngress, Boston, MA, 2017.
17. Anthony T S Ho and Shujun Li. *Handbook of Digital Forensics of Multimedia Data and Devices.* Wiley, Hoboken, NJ, 2015.
18. Ric Messier. *Operating System Forensics.* Elsevier Science, Amsterdam, 2015.
19. Bruce Nikkel. *Practical Forensic Imaging: Securing Digital Evidence with Linux Tools.* Starch Press, San Francisco, CA, 2016.
20. Joel Reardon. *Secure Data Deletion.* Information Security and Cryptography. Springer International Publishing, New York, 2016.
21. Niranjan Reddy. *Practical Cyber Forensics: An Incident-Based Approach to Forensic Investigations.* Apress, New York, 2019.
22. Mark E Russinovich, David A Solomon, and Alex Ionescu. *Windows Internals.* Number pt. 2 in Developer Reference. Pearson Education, London, 2012.
23. Avi Silberschatz, Peter B Galvin, and Greg Gagne. *Operating System Concepts.* Wiley, Hoboken, NJ, 2005.
24. Tyler J Speed. *Asset Protection through Security Awareness.* CRC Press, Boca Raton, FL, 2016.
25. Thomas Sterling, Matthew Anderson, and Maciej Brodowicz. Chapter 18: File systems. In: Thomas Sterling, Matthew Anderson, and Maciej Brodowicz (editors), *High Performance Computing,* pp. 549–578. Morgan Kaufmann, Boston, MA, 2018.
26. Thomas Göbel, Jan Türr, and Harald Baier. Revisiting data hiding techniques for apple file system. In Proceedings of the 14th International Conference on Availability, Reliability and Security, pp. 1–10, Canterbury CA, United Kingdom. 2019.

7
Digital Forensics Requirements and Tools

7.1 Introduction

Previous chapters covered the basics of hard drives and file systems, focusing on how they store data and tools to recover deleted files and forensic images of hard drives. A forensic investigator needs to be familiar with these concepts to produce acceptable results when engaged in a criminal investigation effectively. Furthermore, understanding the inner mechanics of the digital forensic tools employed during an investigation is critical.

An investigator must explain how the digital forensics tools obtained the results with no room for doubt (especially when introducing such evidence in a court of law). But investigators need to trust the tools at their disposal fully. In this chapter, we introduce several forensic tools and their credibility. Furthermore, we will discuss how to evaluate the requirements of such tools in an investigation.

The main objectives of this chapter are as follows:
- To understand how to evaluate the need for digital forensics tools
- To discuss digital forensics techniques and tools
- To further discuss data acquisition tools
- To practically apply various data validation and acquisition techniques and methods using Python and WinHex

7.2 Computer Forensic Requirements

Forensic investigators need to have a keen grasp of several computer-related concepts before they can be a part of a digital forensic investigation [24]. First, investigators need to be familiar with the hardware, including the internal and external devices, the motherboards and various chipsets of hard drives and their settings, and the memory and power connections that make up a computer. The next concept that needs to be understood is the booting process, emphasizing the settings and differences between BIOS and UEFI.

DOI: 10.1201/9781003278962-7

As different operating systems are available and may be installed in suspect computers, their differences and characteristics also need to be understood, with popular versions being Windows (3.1/95/98/ME/NT/2000/2003/XP), DOS, UNIX, LINUX, and Mac. Familiarity with popular software packages like MS Office is also important. Finally, mastery of computer forensic techniques and the software packages that can be employed is crucial.

7.3 Evaluating Needs for Digital Forensics Tools

One of the first considerations that an investigator needs to address is selecting an open or closed source tool. In practice, open-source software is often preferred over closed-source, as it can be easily reviewed by the end-user (forensic examiner) [13]. After selecting an open or closed-source tool, the examiner must ask several important questions about its characteristics. These questions are as follows:

1. What are the operating systems that support the forensic tool?
2. What is the range of file systems that the forensic tool is compatible with and can analyze?
3. Is it possible to automate repeating processes through the use of scripting languages? (e.g., Python, Bash in Linux, and PowerShell in Windows)
4. Is the forensic tool equipped with any automated features?
5. Is the vendor of the forensic tool known for providing adequate support for their tool? What is the vendor's reputation?

Questions 1–2 address the underlying requirements of a digital forensic lab environment and the range of systems that the tool can be used to analyze, as not all forensic tools can be applied to investigate all file systems and file types. The following two questions (3–4) focus on ways of automating the forensic extraction, examination, and analysis phases to avoid unnecessary errors by the examiners due to the repetitive nature of these processes as well as speed up the examination process. Finally, question 5 considers the reputation of the software's vendor and their support capabilities, which can impact the quality of the produced examination results and reports and how a court of law receives them.

7.3.1 Types of Digital Forensics Tools

Digital forensics tools can be broadly categorized into hardware and software. Hardware tools range from single-purpose components like write

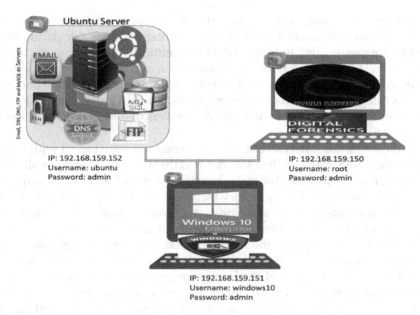

FIGURE 7.1
Three virtual machines included the digital forensics tools used in this book.

blockers [17] that prevent write actions to suspect hard drive to complete computer systems and servers. On the other hand, software forensics tools vary from command-line applications to GUI applications with examples like FTK and Autopsy [3,6]. The tools that are used in the book are displayed in Figure 7.1. Forensic software tools in the Ubuntu Server and Windows 10 VM include the nmap [1] command that can be used to detect open services (ports) on a specified (IP) machine and the fdisk [2] and GParted [4] commands for viewing information about partitions in Windows and Linux.

7.3.2 Tasks Performed by Digital Forensics Tools

Employing trusted digital forensic tools is crucial, as it further supports the validity of collected evidence. Forensic experts are often prompted to follow guidelines, like the ones proposed by NIST's Computer Forensics Tools Testing program [7] for testing forensic tools or the ISO standard 27037 [26] that asserts that Digital Evidence First Responders should use validated tools. Generally, a digital forensic investigation, and thus the actions performed by a digital forensic tool, can be viewed in five major categories: acquisition, validation and verification, extraction, reconstruction, and reporting.

In acquisition, forensic tools are tasked with generating copies of original drives for later analysis. Some sub-functions are physical data copy, logical data copy, data acquisition format, command-line acquisition, GUI acquisition and remote, and live and memory acquisitions. In the validation and verification categories, tools are tasked with confirming that the function as intended (for validation) and employing methods, such as calculating hash values, for proving that two sets of data are identical. A related process to verification is "filtering and analysis," where findings are first collected and then scanned to separate suspicious data from normal data. Some sub-functions of this category include hashing using MD5 and SHA1 algorithms, filtering based on hash values, and analysing file headers to discriminate files based on their file type.

In the extraction/analysis categories, forensic tools are tasked with recovering data from acquired data sources. It is the most challenging of all tasks to master as an investigator and precedes the actual analysis of acquired data. Sub-functions of this category include data viewing, keyword searching, which speeds up the analysis process, decompression, file carving where files are sought in a partition/volume without using a file system, decryption, and bookmarking. In the reconstruction category, tools recreate a suspect drive to make evident what occurred during a crime or an incident.

Usually, this is accomplished by copying an image, often a direct disk-to-image copy of a drive to another location, such as a partition, a physical disk, or a virtual machine. Examples of disk-to-image tools used are the "dd" command [9] in Linux. Finally, in the reporting category, forensic tools are tasked with producing a report, listing the findings of an investigation that investigators can use to communicate their findings in their final report. Some sub-functions include bookmarking or tagging evidence, log reports, and the report generator itself.

7.3.3 Data Acquisition Tools and Formats

In data acquisition, specific file formats are often preferred for storing collected data. These formats include (1) Raw (dd), which refers to disk dump and can be used in Windows and Linux; (2) SMART for Linux, which stores the metadata in separate text files to make them easier to view; (3) E01 (Encase Image File Format) a file format often used by EnCase where hash values and case descriptions are stored in the same file; and (4) AFF (Advanced Forensics Format) which stores all data and metadata in a single file. Several data acquisition tools are available, such as WinHex, FTK imager, and dd and dcfldd in Kali Linux. WinHex allows for generating two types of images, a ".dd" and an application-specific "whx." Although it requires too much time and space, a ".dd" image can be generated and analyzed in Windows or Linux. The FTK Imager tool allows for the generation of forensic images of any of the formats above.

The "dd" [9] and "dcfldd" [8] commands can be used to acquire a forensic image of a partition/device, even allowing for remote acquisition. For example, as depicted in Figure 7.2, we can remotely acquire an image of a partition of the Ubuntu server (192.168.159.152), using "ssh" through the following commands: ssh root@192.168.159.152 "dd if=/dev/sda1" | dd of=server_sda1.

FIGURE 7.2
Example of using dd for data acquisition.

dd bs=512 count=10000 using "dd" and ssh root@192.168.159.152 "dd if=/dev/sda1" | dcfldd of=server_sda_2GB.dd hash=md5 hash=sha1 hashlog=hash_data.txt bs=2048 count=10000 using "ddcfldd."

In the two commands above, the "root@192.168.159.152" is to gain access to the Ubuntu server as the "root" user and the "dd if=/dev/sda1" specifies the input file which is read. After the pipe "|," both "dd" and "dcfldd" define the output file, which is saved in the current directory, through the "of=server_sda1.dd" parameter, and define that each block (unit to be read/written at a time) is "bs" bytes long and "count" number of blocks will be processed. The output file size (for server_sda1.dd and server_sda_2GB.dd) is equal to ((bs*count)/1000(bytes) *1000(Kbytes)), thus for the "dd" example, the file size (in megabytes) is (512*10,000)/(1000*1000)=5,120,000/1,000,000=5.12 MB. The "dcfldd" has similar parameters to the "dd," with the addition of parameters for defining a hashing algorithm through the "hash" parameter and a file to store the produced hash values through "hashlog." The "dcfldd" example defines an output file size of 2 GB (bs=2048, count=10,000), as shown in Figure 7.3.

7.4 Anti-Forensics

The point of forensic investigations is to access a crime scene (whether it is a physical room or a computer system) and identify valuable data. This data can be examined, in the process turning into evidence and allowing for the reconstruction of events before, during, and sometimes after an incident has

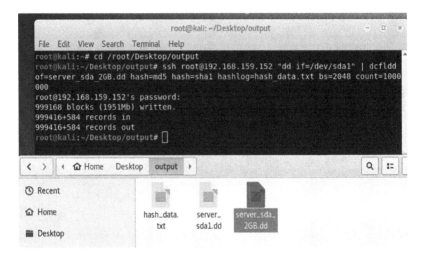

FIGURE 7.3
Example of using dcfldd for data acquisition and preservation.

occurred, leading to the identification of attackers, their methods, and targets. However, the work of forensic investigators is often hindered by these attackers, who seek to cover their tracks by employing a range of techniques called Anti-Forensics [25].

Anti-Forensic software, which can be in place before the acquisition, is tasked with limiting and/or corrupting data that could be used as evidence, performing data hiding and distortion, or even exploiting limitations of known and often used forensic tools. Examples of such software can be found in Metasploit's Anti-Forensic Toolkit [20], like the Timestomp tool, which enables the deletion or modification of NTFS modified, accessed, created, or entry modified timestamp values. The Slacker tool allows the hiding of data in the slack space of NTFS partitions, and the Sam Juicer tool that extracts password hashes from a Windows machine without leaving traces on the disk or registry.

7.5 Evidence Processing Guidelines

Investigators have to adhere to specific rules when handling evidence to avoid bringing the integrity of the evidence to question. Before any data collection can occur, a chain of custody needs to be established, recording who came into contact with the evidence to ensure that it has not been tampered with and establishing a timeline. Guidelines often cover the chain of custody process, with one example being the 16 steps recommmended by

TABLE 7.1

The 16 Steps Recommended by New Technologies Inc. for Processing Evidence

Step Number	Step Name	Step Description
1	Shutdown computer	Considerations: volatile information, remote access, and destruction of evidence
2	Document the hardware configuration of the system	Rigorous notes about configuration before relocation
3	Transport the computer system to a secure location	Do not leave machine unattended
4	Make bit stream Backups of hard Disks and floppy disks	–
5	Mathematically authenticate data on all storage devices	This proves that the investigator has not altered the content of the machine after receiving it
6	Document the system date and time	–
7	Make a list of key search words	–
8	Evaluate swap files	–
9	Evaluate slack space	–
10	Evaluate unallocated space (erased files)	–
11	Search files, file slack, and unallocated space for keywords	–
12	Document file names, dates, and times	–
13	Identify file, program, and storage anomalies	–
14	Evaluate program functionality	–
15	Document your findings	–
16	Retain copies of software used	–

New Technologies Inc. [21] for investigators to adhere to when processing evidence as shown in Table 7.1.

7.6 Implementation of Data Validation and Acquisition Phases

This section explains how to apply validation and authentication using python scripts and Linux terminal commands. Furthermore, data acquisition and data analysing using the WinHex tool are described.

7.6.1 Hash Functions

A hush function is defined as a one-way function that receives data as input of arbitrary length and maps it to an output of fixed length [16]. It is fast and simple to calculate its output f(x) as a one-way function using the input *x*. Still, it is computationally costly (requiring hours of supercomputer processing power) to obtain *x* from *f(x)*. A hush function's output is called a hash, hash value, message digest, or checksum. Although generally, each input will produce a unique output, depending on the algorithm, there is a possibility of a collision occurring [14]. Different inputs will have the same output, which the functions' mathematical theory can explain.

Two commonly used hash functions are MD5, and SHA as briefly explained below:

- **MD5**: The Message-Digest algorithm [23] produces a 128-bit has value and finds use in data integrity checks. However, it is not suited for other fields due to its security vulnerabilities [22,27].

- **SHA**: The Secure Hashing Algorithm family [18] was originally designed by the US's NSA. These algorithms are widely used in cryptographic applications to this day. Depending on the version of the SHA (sha1, sha2, sha3), the digest length ranges from 160 bits to 512 bits. Although they are more robust than the MD5 algorithm, collision attacks have been applied to SHA algorithms [12,15].

For example, to convert the string "Hello World" using the SHA1 function, the output is the following hexadecimal "0a4d55a8d778e5022fab701977c5d84 0bbc486d0," as shown in Figure 7.4.

7.6.2 Authentication and Validation in Digital Forensics

In a digital forensic investigation, two of the most critical actions needed are the authentication and validation of collected data. These steps ensure that the data collected from the crime scene has not been modified, thus ensuring

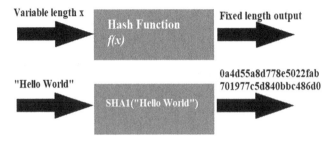

FIGURE 7.4
Example of SHA1 output.

the authenticity of data and lending credibility to the investigation results. The process consists of three steps/rules that need to be observed. First, the collected data (evidence source, like hard disks, partitions, etc.) are hashed, and then the produced digest is stored for later use. Then the image obtained from the collected data is hashed. Finally, the collected data (evidence source) is hashed again, and the produced digest is compared to the one acquired during the first step, with any differences indicating changes to the data.

7.6.2.1 Python Scripts for Hashing

In this section, we will be looking at how to use hash functions, on files, with Python [10]. For the development of the python scripts that we will use to produce the hash digests of files, we will use the Spyder [5] Integrated Development Environment and the module hashlib [11]. On the Ubuntu server, launch the Spyder Integrated Development Environment and create a new file, named "Hashing.py," where our code will be saved. To view the hashing functions available through hashlib, use the following code, shown in Figure 7.5.

When the code is executed, it prints the available hash functions. We focus on the MD5 and SHA1 hashing algorithms, most commonly used in Digital Forensics [19].

7.6.2.2 MD5

First, we will calculate an example of an MD5 hash for the string "Digital Forensics," given by the following code in Figure 7.6. In the code, "hash_md5"

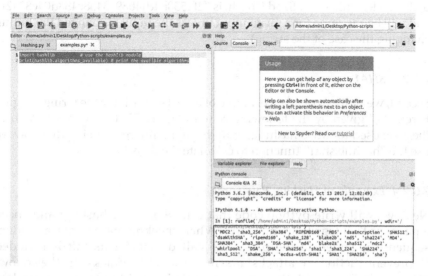

FIGURE 7.5
Available hashing functions from hashlib.

FIGURE 7.6
Example of Python code for calculating an MD5 hash.

FIGURE 7.7
Example of code for SHA1 hash calculation.

is a variable that stores the hash value of the "Digital Forensics" string and the "encode()" function converts the string into byte format, as the "hashlib. md5" function receives a sequence of bytes as input in Python 3. The output of this code, which is of fixed length, is "2f2550b4df4b89e3438c438d04c83a21" and will be the same for a given input unless the input is changed (including changing the case of the string).

7.6.2.3 SHA1

Second, we will calculate an example of an SHA1 hash for the string "Digital Forensics," given by the following code in Figure 7.7. The code is similar to the code used for MD5 hashing, with the main difference being that now we use the "hashlib.sha1" function to calculate the digest.

7.6.2.4 Example of Hashing Passwords

Next, we will write python code that can be used for hashing and check a password's integrity using SHA1. When checking passwords, a random sequence is added to the password, called "salt." This salt is included to prevent dictionary attacks and rainbow table attacks. In the following code in Figure 7.8, the "uuid" module is used to generate a random number.

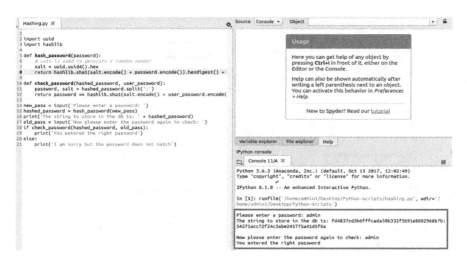

FIGURE 7.8
Example for password hash calculation and its validity.

7.6.3 Hashing and Data Acquisition

The process of checking the integrity of data collected during the data acquisition process of a digital forensic investigation will be similar to the one presented in the previous hashing password section, with the hashes compared between the original files (original evidence) and copied files (images). In the Ubuntu terminal, gain superuser privileges by typing "sudi -i" and moving to "cd /home/admin1/Desktop." Create a file named "hashing.txt" and add the following string "Digital Forensics" in the file.

To calculate the hash digests, use "echo –n" use the "md5sum," and sha-1sum piping the output of the echo command by typing "echo -n hashing.txt | md5sum" for MD5 and "echo -n hashing.txt | sha1sum" for SHA1. If the file is not modified, consecutive executions of the above code will produce the same hash digests for each hash function.

7.6.3.1 Data Acquisition Using WinHexs

We will check the integrity of a copied partition in Windows 10 VM. First, we will create a copy of partition (H). Launch WinHex with administrator privileges (right-click, Run as administrator) and navigate to "Tools," "Open Disks," and select "NTFS_Train(H)." Next, select "File," "Create Disk Image," as shown in Figure 7.9, and then "Ok." After selecting "Ok," the MD5 hash of the drive will be displayed by WinHex as given in Figure 7.10, and it can also be found at the directory where the image was stored, in a ".txt" file.

Next, open the image that was just extracted with WinHex and calculate its hash by navigating to "Tools," "compute hash," "select MD5 (128 bit)." Then, to validate that the image has not been modified, compare the produced

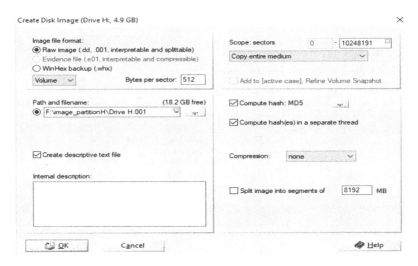

FIGURE 7.9
Creating disk image in WinHex.

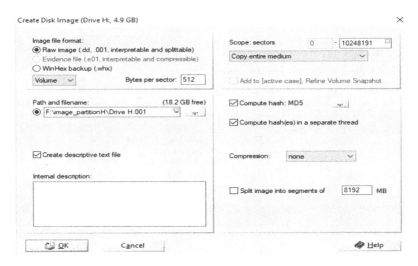

FIGURE 7.10
WinHex-generated MD5.

MD5 hash of the image to the pre-calculated MD5 of the original source (disk drive), stored in the same directory where the image was stored. From this image, it is possible to recover files and directories. By selecting "Specialist" and then "Interpret image file As Disk," a window appears with a list of files and directories of the hard disk image that was opened that can be copied and thus recovered.

7.7 Conclusion

This chapter has discussed several digital forensic tools employed by investigators and the tasks they should perform to follow a legal digital forensics process. Furthermore, we discussed what concepts need to be considered when evaluating digital forensic tool needs and addressed some recommended steps necessary to conduct a forensic investigation soundly. Various data validation and acquisition tools were also explained. In Chapter 8, we will discuss the file allocation table file system.

References

1. Nmap tool, https://linux.die.net/man/1/nmap, 2021.
2. fdisk tool, https://home.csulb.edu/~murdock/fdisk.html, 2021.
3. Autopsy tool, https://www.sleuthkit.org/autopsy/, 2020.
4. gparted tool, https://gparted.org/, 2021.
5. Spyder as a python IDE, https://docs.spyder-ide.org/current/installation.html, 2009.
6. Ftk imager tool, https://accessdata.com/product-download/forensic-toolkit-ftk-version-6-0, 2021.
7. Nist's computer forensics tool testing (cftt) program, https://www.nist.gov/itl/ssd/software-quality-group/computer-forensics-tool-testing-program-cftt, 2021.
8. Dcfldd command, http://rpm.pbone.net/info_idpl_74941372_distro_mageia8_com_dcfldd-1.3.4.1-11.mga8.x86_64.rpm.html, 2021.
9. dd command, https://www.geeksforgeeks.org/dd-command-linux/, 2021.
10. Python, https://www.python.org/, 2021.
11. hashlib, https://docs.python.org/3/library/hashlib.html, 2021.
12. Zeyad Al-Odat, Assad Abbas, and Samee U Khan. Randomness analyses of the secure hash algorithms, sha-1, sha-2 and modified sha. In *2019 International Conference on Frontiers of Information Technology (FIT)*, pp. 316–3165, Islamabad, Pakistan. IEEE, 2019.
13. Cory Altheide and Harlan Carvey. *Digital Forensics with Open Source Tools*. Elsevier Science, Amsterdam, 2011.
14. Jason Andress. Chapter 5: Cryptography. In: Jason Andress (editor), *The Basics of Information Security (Second Edition)*, pp. 69–88. Syngress, Boston, MA, 2014.
15. Karthikeyan Bhargavan and Gaëtan Leurent. Transcript collision attacks: Breaking authentication in TLS, IKE, and SSH. *Computer Science, Mathematics* 2016. DOI: 10.14722/NDSS.2016.23418.
16. Thomas W Edgar and David O Manz. Chapter 2: Science and cyber security. In: Thomas W Edgar and David O Manz (editors), *Research Methods for Cyber Security*, pp. 33–62. Syngress, Boston, MA, 2017.
17. Scott R Ellis. Chapter 40: Cyber forensics. In: John R Vacca (editor), *Computer and Information Security Handbook* (Third Edition), pp. 573–602. Morgan Kaufmann, Boston, MA, 2013.
18. James T. Harmening. Chapter 58: Virtual private networks. In: John R Vacca (editor), *Computer and Information Security Handbook* (Third Edition), pp. 843–856. Morgan Kaufmann, Boston, MA, 2017.
19. Anthony TS Ho and Shujun Li. *Handbook of Digital Forensics of Multimedia Data and Devices*. Wiley, Hoboken, NJ, 2015.

20. Hamid Jahankhani. *Handbook of Electronic Security and Digital Forensics*. World Scientific, Singapore, 2010.
21. Thomas Johansson, Subhamoy Maitra, ICC India, INDOCRYPT, and LINK (Online service). *Progress in Cryptology – INDOCRYPT 2003: 4th International Conference on Cryptology in India*, New Delhi, India, December 8–10, 2003, Number v. 4 in Lecture Notes in Computer Science. Springer, 2003.
22. Vlastimil Klima. Tunnels in hash functions: Md5 collisions within a minute. *IACR Cryptology ePrint Archive*, 2006:105, 2006.
23. Zeyad A Al-Odat. Analyses, mitigation and applications of secure hash algorithms. Doctoral dissertation, North Dakota State University, 2020.
24. Jason Sachowski. *Digital Forensics and Investigations: People, Process, and Technologies to Defend the Enterprise*. Taylor & Francis, Boca Raton, FL, 2018.
25. Shiva VN Parasram. *Digital Forensics with Kali Linux: Perform Data Acquisition, Digital Investigation, and Threat Analysis Using Kali Linux Tools*. Packt Publishing, Birmingham, 2017.
26. ITS Techniques. ISO/IEC 27037: 2012 information technology security techniques guidelines for identification collection acquisition and preservation of digital evidence. ISO/IEC-The standard was published in October, 2012.
27. Xiaoyun Wang and Hongbo Yu. How to Break MD5 and Other Hash Functions. In: Cramer R. (editor) Advances in Cryptology – EUROCRYPT 2005. EUROCRYPT 2005. Lecture Notes in Computer Science, vol 3494. Springer, Berlin, Heidelberg. https://doi.org/10.1007/11426639_2.

8

File Allocation Table (FAT) File System

8.1 Introduction

In previous chapters, we discussed the two main file systems in contemporary Windows computers, the NTFS and FAT. Although we briefly discussed their characteristics, we will further delve into the FAT file system and analyse its structures and FAT partitions in this chapter. Furthermore, we will discuss various methods of FAT partition analysis for digital forensics purposes.

The main objectives of this chapter are as follows:

- Understand FAT FS types and their data structures
- Discuss FAT partition and its files and directories
- Learn different methods of FAT partition analysis
- Apply the FTK imager tool for data acquisition and integrity and the WinHex, Disk Editor, and Autopsy tools for forensics analysis for the FAT file system

8.2 File Allocation Table (FAT)

File Allocation Table (FAT) is one of the simplest file systems to implement and understand, developed by Microsoft for MS-DOS and used to be the primary file system for the consumer versions of Microsoft Windows [6,2,15]. Virtually all existing operating systems support it for personal computers and, although it may not be the default file system for Windows-based computers anymore, it is an ideal format for floppy disks and solid-state memory cards. It is considered a convenient way of sharing data between different operating systems installed on the same computer.

Important components include FAT (which is where the file system gets its name from), where clusters of files are linked to form files (two copies for redundancy FAT1, FAT2), Boot Sector and the Directory Table Entry, where information about each file, and their starting sector are stored. The FAT file system ID is stored in the MBR at offset "0x52" found in "FAT32 Boot Sector" templates, using the "Active @ Disk Editor." The file system ID can also be found in the "Master Boot Record" template and then "Edit Boot Record."

8.2.1 Common Types of FAT

The FAT file system comes in four general types, FAT12, FAT16, FAT32, and exFAT (extended FAT). The numbers 12, 16, and 32 indicate the number of clusters available by the versions of FAT. FAT12 was originally designed for floppy disks and MS-DOS [9,10]. It uses 12-bit addressing for clusters, allowing for 2^{12}=4096 clusters, with four sectors per cluster and usually 512 Bytes for each sector, supporting volumes of up to *clusters* ∗ *sectors*=4096 ∗ 4 ∗ 512=4096 ∗ 2048=8 MB. However, other values were also used (1024 bytes, 4096 bytes) [10].

FAT16 is the next version after FAT12, with 16-bit cluster addresses and 4 sectors per cluster, supporting volumes smaller than 16 MB or up to 128 MB [9], depending on the cluster size. FAT32 employs 32-bit cluster addressing, although only 28 bits are used, thus allowing the addressing of 2^28 clusters, with 8 sectors per cluster (sector sizes options include 1024, 2048, 4096 MB for each sector), with a max volume size of up to 8 TB [2,6]. Finally, the exFAT file system utilizes a cluster bitmap for fast allocation and deletion, supporting a maximum file size of 16 exabytes ($2^{64}-1$ bytes) and a max volume size of 128 petabytes [9]. It includes a single bit as an indicator for files stored in a contiguous way, having a better on-disk layout, although exFAT is not compatible with all operating systems [16].

8.2.2 FAT Layout

The layout of FAT12/16 and FAT32 are similar, as can be seen in Figure 8.1. All three versions of FAT share four common layout elements, the Reserved Area,

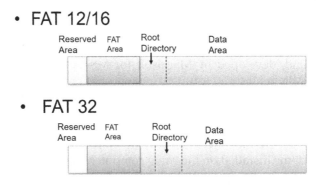

FIGURE 8.1

Layout of FAT12/16 and FAT32.

TABLE 8.1

Elements of a FAT Volume

FAT Elements	Description
Boot Sector	Contains information about the disk and code that is loaded and run by the BIOS/UEFI start-up
FAT1	Holds information about what clusters are allocated to files on the disk
FAT2	Copy of FAT1 kept for redundancy and security
Root directory	Holds information about the directories and files, including the first cluster of each file
Data area	Stores the contents of files and directories

FAT are (FAT1, FAT2), the Root Directory, and Data area [6]. The main difference between the layouts of FAT12/16 and FAT32 is that in FAT12/16 the Root Directory is always located at the beginning of the Data Area, while in FAT32, it can be located anywhere within the same (Data) area. In addition, the first 16 bytes for the three FAT versions include the common attributes, JMP Instruction, OEM ID, Bytes per sector, Sectors per cluster, Number of FATs. The common elements found in a FAT Volume/Partition are further described in Table 8.1.

8.3 FAT Layout Analysis

The first step in analysing a FAT file system is to identify the physical layout areas of the file system. The first element of FAT to be determined is the reserved area, which starts at volume sector 0, and its size (in sectors) is given in the Boot Sector [5]. The reserved area for FAT12/16 is 1 sector, while FAT32 reserves 4 sectors (Boot Sector, Boot Sector Copy, and two more sectors), all ending with the signature "0x55 AA." The FAT area begins after the reserved area, with its size equal to the number of FAT structures multiplied by the size of each FAT.

The data area begins after the FAT area, and it includes the Root Directory. As it is the final element of a FAT file system, the size of the data area is determined by subtracting the starting address of the data area from the total number of sectors in the file system.

As seen in Figure 8.2, the FAT area includes two tables, FAT1 and FAT2 (copy of FAT1), that store information about what clusters of the partition are allocated, with one entry existing in the FAT table for each cluster in the partition. The size of the FAT table is pre-determined, which means that the number of clusters in a FAT file system is fixed. According to the status of a cluster, its entry in the FAT tables may store different values. Entries of unallocated clusters (free space) are indicated by the "0x0" value.

Reserved   FAT area   Data Area
Sector 0

FIGURE 8.2
Location of FAT is in FAT volume.

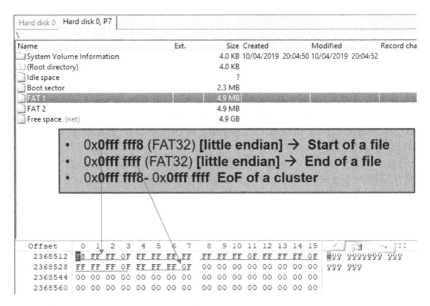

FIGURE 8.3
Example of analysing FAT Hex code.

Clusters that are identified as damaged (bad clusters) contain a bad cluster marker, which is the reserved hexadecimal value "0xff7" for FAT12, "0xfff7" for FAT16, and "0x0fff fff7" for FAT32, with the hexadecimal values represented in big-endian form. Remember that FAT uses little-endian representation, as such a bad cluster in FAT16 would be indicated by the value "0xf7ff." If clusters are in use, the marker for the starting cluster of a file is "0xff8" for FAT12, "0xfff8" for FAT16, and "0x0fff fff8" for FAT32, and the last cluster of a file contains the end of file, "0x0ff" for FAT12, "0x0fff" for FAT16, and "0x0fff ffff" for FAT32. As such, a file in FAT32 would begin with "0x0fff ffff8" and end with "0x0fff ffff," as shown in Figure 8.3.

If a FAT entry does not have one of the aforementioned reserved values, it stores the logical address of the next cluster that belongs to the file. Files may be fragmented if they are not stored in contagious clusters, with the clusters linked through the FAT entries (entries store location of next cluster). Creating files or directories results in the allocation of clusters for storing

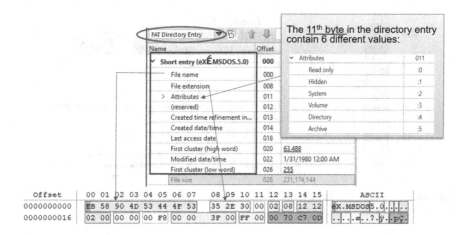

FIGURE 8.4
Explanation of FAT directory entry using active disk tool.

their content and new directory entries that store the address (number) of their first cluster.

In FAT, creating a directory results in the allocation of a cluster, with its bits set to zero [2,14]. As shown in Figure 8.4, each directory entry is stored in 32 bytes, with the 11th byte called "Attributes" and used to indicate the type of directory entry (8 values possible) like read-only, archive file, hidden file, system file, and subdirectory. A 512-byte sector can store 16 directory entries (512/32). The size attribute of a directory in the directory entries is always set to 0. However, its size can be determined by going to its starting cluster and following the cluster chain until the EoF is reached (reserved hexadecimal value). A set of directory entries represent the contents of a directory, with a new entry allocated when a file is created.

Directory entries are stored in at least one cluster, with the first two directory entries of a directory representing the "current directory (.)" and the "parent directory (..)." To determine the creation time of a directory, the time fields can be used; however, the last written date cannot be confirmed, as the directory entries are not updated for each directory modification. To access a file or directory in FAT, first, its directory entry is accessed to determine the first cluster of its content. Then, the next cluster is identified, as each cluster stores the address of the next cluster in the cluster chain that forms the file, with the process continued until the EoF code is reached ("0x0fff ffff" for FAT32). The process can be seen in Figure 8.5.

The status of the directory entry, and thus of the file it represents, is given in the first byte of the entry [6]. A deleted file starts with the "0XE5" hex value, marking the entry as reusable, while the end of directory entries in a cluster is represented by "0x00." As the first directory entry of any directory

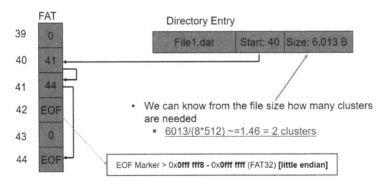

FIGURE 8.5
Cluster chain for accessing files in FAT.

is the ".", which references the current directory, all directories share the following hexadecimal header "0x2E 20 20 20 20 20 20 20 20 20 20 20," where "0x2E" is "." and "0x20" is an empty space, as shown in Figure 8.6.

The next byte after this header (offset "11") indicates the type of entry, with "0x10" representing any directory and "0x08" referring to the volume label which resides in the root directory. A representation of the FAT layout and structure can be seen in Figure 8.7. The figure displays the process of accessing the contents of a file by first identifying "dir1" in the root directory and then following the cluster chain until the EoF is reached.

An important drawback of most FAT implementations is fragmentation [8]. Fragmentation occurs when a file cluster is nulled and then reused to store a file that cannot be stored in a single cluster. This results in the contents of a file being scattered around the drive, which hinders reading/writing speeds. A technique to counter this issue is "defragmentation," which results in files stored in consecutive sectors, although the process is lengthy and needs to be applied regularly to maintain the high performance for FAT drives.

8.3.1 FAT Analysis

When tasked with analysing FAT partitions, several areas need to be investigated for hidden data [7,12]. First, clusters that are marked as "bad" ("0x0FFFFFF7" for FAT32) may hide information and need to be checked, as the OS ignores them. Deleted files and directories are also potential locations for hiding data. The allocated size of directories and the number of files they contain need to be considered as well, as data may be hidden in a directory's unused allocated space.

Furthermore, the slack space needs to be investigated, as the volume size may not be multiple of the cluster size, resulting in some sectors being unused. Other spaces that may hide data include the space between the last entry in

- 0x**00** → This directory entry is not used.
- 0x**E5** → (File deleted) Free to be REUSED this Directory Entry

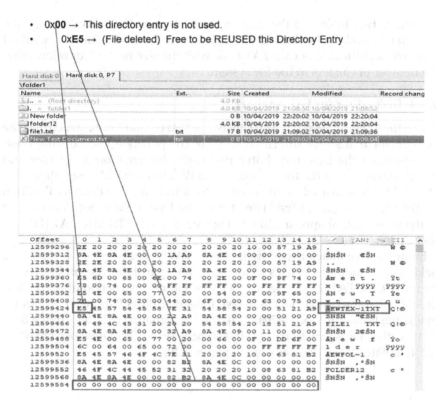

FIGURE 8.6
Deleted and free directory entries in FAT.

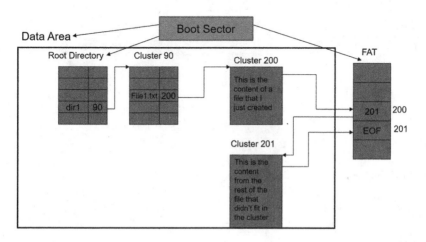

FIGURE 8.7
Accessing file in FAT.

a primary FAT table and the start of the backup or the space between the backup FAT and the beginning of the data area. A suitable analysis method is to compare the size of each FAT table with the size needed for referencing the number of clusters in the file system.

8.3.2 Disk Editor for FAT Analysis

Recall that a FAT file system [15] consists of two primary data structures, the Boot Sector that stores information about the partition used during boot-ing time and the Directory Entry that stores information about files and directories stored in the file system. In the Windows 10 VM, you will use the Disk Editor tool to analyse the Boot Sector and Directory Entry of Partition (I:) (FAT32). Navigate to the "File" tab in the Disk Editor tool, "open," and in the window that appears, like in Figure 8.8, select "TRAIN_FAT (I:)" and then "open."

From the Templates dropdown, select "FAT32 Boot Sector" to view the fields that Windows OS uses to load this FAT partition. As can be seen in Figure 8.9, the FAT32 Boot Sector includes a jump instruction (first 3 bytes), the BIOS Parameter Blocks (53 bytes), Extended BIOS Parameter Blocks (26 bytes), the Bootstrap Code (420 bytes), and the signature (2 bytes, "0xAA 55"). To view the "Directory Entry" (32 bytes each), select "Fat Directory Entry" in the Templates dropdown, as seen in Figure 8.10.

FIGURE 8.8
Open TRAIN_FAT (I:) in disk editor.

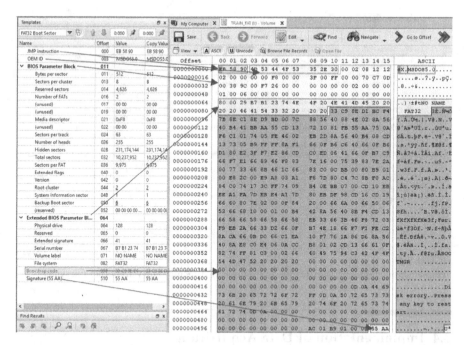

FIGURE 8.9
FAT32 Boot Sector values of TRAIN_FAT (I:).

FIGURE 8.10
FAT32 values of TRAIN_FAT (I:).

8.3.3 WinHex Tool for FAT Analysis

Start WinHex with administrator privileges (right-click, "Run as Administrator"). Go to "Tools," "Open Disk" and select "HD0: VMware, VMware Virtual s (120 GB, SAS)." Then, select "Partition 7 (I:) and then "View," "Template Manager." In the template manager, to view the boot sector, select "FAT 32 Boot Sector" and click "Apply!," as shown in Figure 8.11. To view the directory entry, in the template manager, select "FAT Directory Entry" and click "Apply!." To analyse the FAT32 partition (I:), in the WinHex tool, right-click on partition (I:) and select "Explore."

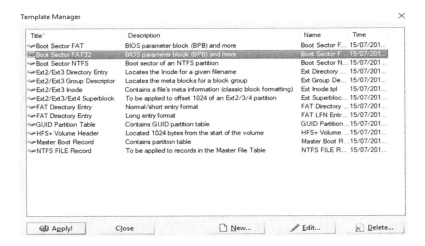

FIGURE 8.11
Template manager using WinHex for analysing FAT directory entry and other file systems.

8.4 Implementation of Data Acquisition and Analysis in Windows

The data acquisition process using the FTK imager tool in Windows 10 is explained. Moreover, analysing the produced images using the Autopsy tool is discussed.

8.4.1 Prerequisites for Doing These Exercises

Do the following steps:

1. Launch the Windows 10 VM.
2. Format Partition (H:) and Partition (I)
3. Right-click on Partition (H:) and select Format
4. Right-click on partition (I:) and select Format
5. Open Partition (E:) Copy the folders "general files" and "Photos" and Paste them to Partition (H:) and Partition (I:)
6. Delete some files from Partition (H:) and Partition (I:)

8.4.2 Data Acquisition and Analysis of FAT

During the data acquisition phase of a forensic investigation, forensic experts recover potentially useful sources of evidence by extracting a forensic image from collected media, such as hard drives, removable hard drives, USB, CDs/DVDs, and other storage media obtained from servers,

PCs, laptops, gaming consoles and other devices [4]. In the later phases of the investigation, this acquired image is used for extracting and analysing evidence rather than the original source, with the forensic image verified against the original to ensure that the former is an exact duplicate of the latter.

A disk image is a file that replicates the contents, structure, and metadata found in a storage medium, effectively duplicating the original [11]. Although it might sound similar to a normal backup, a forensic disk image maintains the integrity of the exact storage structures of the original, which is of utmost importance as it ensures the integrity of a forensic investigation. Suppose it cannot be asserted that a disk image is an exact duplicate (content and structure) of the original data source. In that case, the integrity of any identified evidence is in jeopardy and could be inadmissible in a court of law [13,17]. As such, acquiring a forensic disk image file from a target device is the first step of any digital forensic investigation.

8.4.2.1 The FTK Imager Tool

The Forensic Toolkit (FTK) tool [3] is imaging software created by AccessData, capable of producing local and remote images. The free version enables sonly local imaging and can acquire images of storage devices such as hard disks, USB, CD drives, and individual files. To start the FTK imager, in the Windows 10 VM, navigate to the "Access Data FTK Imager" in the Desktop. A screen will appear, similar to Figure 8.12.

To create a disk image, navigate to "File" and select "Create Disk Image." Options allow the capture from memory and imaging of individual items. Next, you are prompted to select the Source Evidence Type, with "Logical Drive" indicating that a portion of the hard disk will be imaged, while a

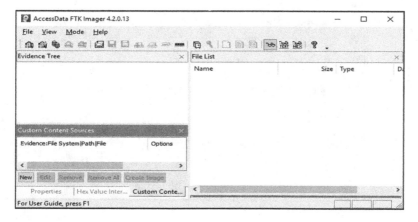

FIGURE 8.12
The FTK imaging tool.

"Physical Drive" produces the image of the entire hard disk. Then, a window with a list of available drives will be provided, allowing for the selection of the volume to be imaged. In this case, select "H:\-Train_NTFS" and then click "Finish."

A "Create Image" window will appear, displaying the image source (volume to be imaged). Click the "Add" button to select the image type and the destination of the image file that will be generated. As shown in Figure 8.13, the image type refers to the kind of image that will be produced, with four possible choices, Raw (dd) that refers to "disk dump" that can be used in Windows and Linux, SMART that stores the metadata in a separate text file for easier viewing, E01 (our choice for this lab) that stores hash values, case description information and other details in the same file (used by EnCase an enterprise digital forensics program) and AFF that stores all data and metadata in a single file.

After selecting the type of image and output, a window will prompt you to add information about the case for which this image will be produced. This option is a great help during an investigation, as keeping track of evidence and having detailed notes is extremely important. For this lab, enter the following: Case Number: "1," Evidence Number: "1," Unique Description: "Imaging H Drive on my local hard disk," Examiner: your name. The folder where the imaging file will be stored is "F:\My-Images" and give the file a suitable name, like "H Drive-Local Disk." The "Image Fragment Size" allows you to specify the size (in megabytes) of the fragments in which the image file will be split. This option is ideal when working with large disk images. If the value given is greater than the size of the hard disk, then only one file is created with the original hard disk's size.

FIGURE 8.13
Image types using the FTK imager.

The next option specifies compression of the image, although some image types (like "dd") cannot be compressed, while others (like "E01") can. After finishing with this window, make sure that the "Verify images after they are created" is selected, as it allows for the automatic creation of hash values, which will later be used to verify the image's validity. After the imaging process is completed, an "Image Verify Results" window will appear, displaying both MD5 and SHA1 hash values of the image. Storing these hash values is important, as they should be reproducible after the analysis to assert the integrity of the data. Selecting the "Image Summary" option displays a detailed summary of the generated image, including the produced hash values. In addition, this information has been copied to a text document located in the same directory as the disk image.

8.4.2.2 The Autopsy Tool

Autopsy [1] is an open-source digital forensics platform that can be used to analyse data sources like hard drives and mobile devices and recover evidence from them. In this section, we will be analysing data with Autopsy. Data analysis with Autopsy requires several steps to be taken.

First, a "Case" is created, with the case being a container for one or more data sources. Next, the data sources are added to the case, followed by their analysis through Ingest modules that work in the background. Then, the investigator manually navigates Autopsy's interface, viewing file content and ingesting module results to identify evidence, which can be tagged for secondary analysis and reporting. Finally, the investigator can initiate a report generation based on tagged evidence and results obtained through analysis.

We will use Autopsy to analyse the image file that we previously generated through FTK. After launching Autopsy by double-clicking its desktop icon, you first need to create a new case in which the disk image will be opened for analysis. Each case has its own directory, configuration files, a database, reports, and other generated files. A case can include data sources from multiple drives of the same computer or multiple computers. To start, select "Create New Case," and in the "New Case Information" window that appears, enter the case name "case1" and base directory "C:\data."

After clicking next, you will need to add more information like the case number and investigator's name (enter "1" and your name, respectively). After entering the requested information (and clicking "Finis"), a minute is required to create the database for the case and all the necessary modules. Next, select "Disk Image or VM file as Data Source," as shown in Figure 8.14, to load the E01 file that FTK generated.

You will then be prompted to enter the path to the data source. The disk image was stored in the "F" drive, with its full path being "F:\data-acquisition of E drive\img1.E01." The time-zone should be "GMT +10:00 Australia/ Sydney," as that value was used when the image was created. However, the time-zone information does not affect the metadata of the image and can

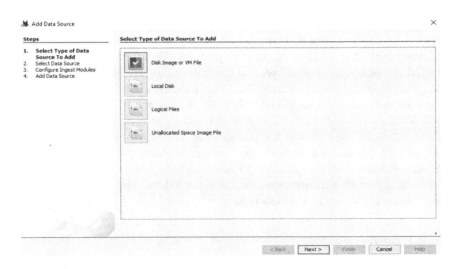

FIGURE 8.14
Selecting data source in Autopsy.

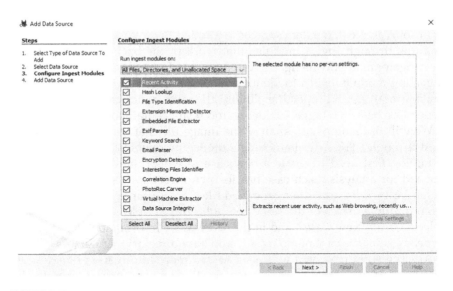

FIGURE 8.15
Selecting ingest modules in Autopsy.

be set to "GMT +0:00" if the investigator is uncertain. The next step is to ingest module selection. Ingest modules perform analysis on the selected data source, parting they contend, calculating hash values, and searching for keywords. Navigate to "Select All" and then "Next," as shown in Figure 8.15.

After clicking "Finish," a new window will appear, displaying the data source, file types, and deleted files, including their metadata (in "Views")

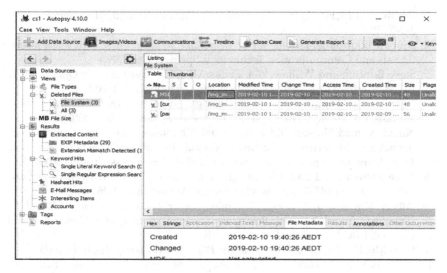

FIGURE 8.16
Analysis and results of a data source in Autopsy.

and ingest module results like email messages and EXIF files, as shown in Figure 8.16. You can browse the Data Sources and their files. In Views, you can find Deleted Files and their Metadata. In Results, you can discover EXIF Metadata and their Hex and Metadata, etc.

8.5 Conclusion

In this chapter, we discussed the elements of the FAT file system, including the Reserved, FAT, and Data areas. Furthermore, the analysis of a FAT partition was explained, which can be used for identifying evidence in FAT directories. The FTK imager tool was used for data acquisition and integrity for the FAT file system. The WinHex, Disk Editor, and Autopsy tools were employed to analyse the FAT file systems to find evidence. Next, the NTFS file system will be explained.

References

1. Autopsy tool, https://www.autopsy.com/, 2021.
2. Ec-Council. *Computer Forensics: Investigating File and Operating Systems, Wireless Networks, and Storage (CHFI)*. Computer Hacking Forensic Investigator. Cengage Learning, Boston, MA, 2016.

3. Ftk imager tool, https://accessdata.com/product-download/ftk-imager-version-3-2-0, 2017.
4. Ahmed Bouridane. *Imaging for Forensics and Security: From Theory to Practice.* Signals and Communication Technology. Springer US, 2009.
5. Steve Bunting and William Wei. *EnCase Computer Forensics: The Official EnCE: EnCase?Certified Examiner Study Guide.* Wiley, Hoboken, NJ, 2006.
6. Brian Carrier. Digital crime scene investigation process. In: *File System Forensic Analysis*, pp. 13–14. Addison-Wesley Professional, Boston, MA, 2005.
7. Nihad Ahmad Hassan and Rami Hijazi. Chapter 4: Data hiding under windowsÂ R os file structure. In: Nihad Ahmad Hassan and Rami Hijazi (editors), *Data Hiding Techniques in Windows OS*, pp. 97–132. Syngress, Boston, MA, 2017.
8. Dave Kleiman and Laura E Hunter. *Winternals Defragmentation, Recovery, and Administration Field Guide.* Elsevier Science, Amsterdam, 2006.
9. Albert Marcella and Doug Menendez. *Cyber Forensics: A Field Manual for Collecting, Examining, and Preserving Evidence of Computer Crimes* (Second Edition). CRC Press, Boca Raton, FL, 2010.
10. Scott Mueller. *Upgrading and Repairing PCs.* Que, Indianapolis, IND, 2003.
11. Bruce Nikkel. *Practical Forensic Imaging: Securing Digital Evidence with Linux Tools.* Starch Press, San Francisco, CA, 2016.
12. Tammy Noergaard. Chapter 5: File systems. In: Tammy Noergaard (editor), *Demystifying Embedded Systems Middleware*, pp. 191–253. Newnes, Burlington, MA, 2010.
13. Niranjan Reddy. *Practical Cyber Forensics: An Incident-Based Approach to Forensic Investigations.* Apress, New York, 2019.
14. Richard E Smith. *Elementary Information Security.* Jones & Bartlett Learning, Burlington, MA, 2015.
15. Thomas Sterling, Matthew Anderson, and Maciej Brodowicz. Chapter 18: File systems. In: Thomas Sterling, Matthew Anderson, and Maciej Brodowicz (editors), *High Performance Computing*, pp. 549–578. Morgan Kaufmann, Boston, MA, 2018.
16. Jeffrey R Shapiro, Jim Boyce, and Rob Tidrow. *Windows 10 Bible.* Wiley, Hoboken, NJ, 2015.
17. John R Vacca. *Managing Information Security.* Elsevier Science, Amsterdam, 2013.

9

NTFS File System

DOI: 10.1201/9781003278962-9

9.1 Introduction

In Chapter 8, we discussed the FAT file system, one of the two most famous contemporary file systems that are found in Windows machines. In this chapter, we will discuss the NTFS file system and its elements in more detail. In addition, we will explain the available methods that can be employed to analyse NTFS volumes.

The main objectives of this chapter are as follows:

- Discuss elements of the NTFS file system
- Learn NTFS reserved files and directories
- Understand NTFS structure for holding data
- Apply digital forensics analysis and implications for NTFS volumes using the Disk Editor, Winhex, and Autospys tools to find evidence and produce forensics reports

9.2 New Technology File System (NTFS)

NTFS is a file system developed by Microsoft and was intended to overcome the limitations of the FAT file system and replace it. NTFS manages the storage, organization, and retrieval of files in a hard disk volume like other file systems. It is the default file system for Windows OS starting from 1993 with Windows 3.1 [3,15]. The benefits of NTFS compared to other file systems are as follows:

- Its increased reliability is a journaling file system and thus maintains a log of changes made to the data. This allows it to "roll back" and re-initiate actions that may be interrupted by unexpected circumstances such as power outages.

- It provides expanded security features, such as permission restrictions, ownership, encryption of files, and access controls. This enables the administrator to allow certain users to access appropriate files and thus maintain the confidentiality of sensitive files.
- It supports large storage media, with its performance not diminishing as the drive's size increases.
- Additionally, it enables the creation of very large files, up to 18.4 Exabytes. This is due to the fact that NTFS utilizes scalable and generic data structures.

However, NTFS also has some disadvantages, including:
- It requires high space "overhead" for storing file system files, and as such, it is not suitable for smaller storage media.
- It is ineffective for cloud environments.
- It is not a native copy-on-write file system, which would allow it to share a copy of a file that has not been modified yet, between processes that access it. This allows the file system to conserve space by providing pointers to a file until a modification is made, a copy is created.

NTFS utilizes the ARIES algorithm to perform journaling tasks when data is created or modified [4]. The ARIS algorithm is a write-ahead logging technique, which ensures that changes are first safely recorded. Then they are applied, thus enforcing that when changes are applied to files, they will be applied in their entirety (atomicity). The effects of those changes will be persistent (durability) [7]. In NTFS' case, any change that will be made on a file's metadata is first logged in a specialized file called "$LogFile" located in the root directory.

After the changes have been recorded to the "$LogFile" file and it has been securely stored in the hard disk, the changes are then applied to the metadata of the intended file. The "$LogFile" functions in a circular mode, where the initial records (called "pages") are deleted when the file reaches maximum capacity, with the space reused for new records. Every 5 seconds, NTFS creates a checkpoint in the log file, which enables it to identify the sequence of commands necessary to recover the volume in the event of a catastrophic crash [13].

9.3 NTFS Architecture

In NTFS, everything is considered to be a file, including the various file system tables, files, and metadata. The file system's metadata are stored as hidden files, with their names often beginning with the character "$" [2]. The

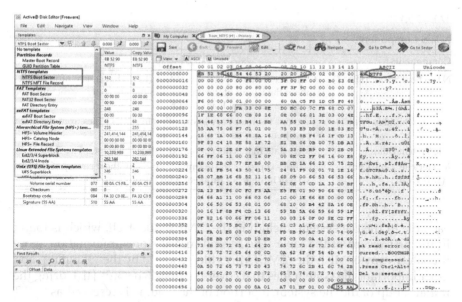

FIGURE 9.1

Example of NTFS Boot Sector hex information using active disk editor.

boot code, which is the code necessary to launch the OS of the volume, is stored in the "$Boot" file and needs to be at the beginning of the volume, as depicted in Figure 9.1. It is important to point out that a copy of the boot sector and the "$Boot" file are often stored at the end of the volume for redundancy, although these copies are not part of the file system allocated space.

All the information about the files that the NTFS manages is stored in the Master File Table in the form of a file named "$MFT" and is integral for the file system's operations [3]. The "$MFT" maintains records for all directories and files that can be found in the record, with the records being two sectors long (1024 Byes) each and additional records assigned to a file if its metadata exceeds the two-sector limit. A copy of the "$MFT," called "$MFTMirr," is maintained in case the original is corrupted, thus allowing the recovery of the contents of the file system. The "$MFTMirr" file is stored in the middle of the allocated space, where the users create and store files. In NTFS, all the available space in the volume is addressable and thus usable. Table 9.1 lists the essential components of the NTFS file system.

9.3.1 Master File Table (MFT)

The Master File Table (MFT) is a core element of NTFS, as it stores important information, in the form of a record, for all files and directories that can be found in the volume. In addition to storing metadata, similar to the directory entry structure found in the FAT file system, MFT also tracks the information about cluster allocation for each file [3]. MFT stores at least one record for

TABLE 9.1

NTFS Components

Components	Description
NTFS Boot Sector ($BooT)	Contains the BIOS parameter block (BPB) that stores information about the layout of the volume and the file system structures and the boot code that loads the OS
Master File Table ($MFT)	Contains the necessary information to retrieve files from the NTFS partition, such as the attributes of a file
File System Data	Stores data that is not contained within the Master File Table
Master File Table Copy ($MFTMirr)	Includes copies of the essential records for the recovery of the file system if there is a problem with the original copy

each file in the file system, including the "$MFT" file itself, which is found in the first record of the table. The second record stores information about the safety copy of MFT, the "$MFTMirr" file, followed by the "$LogFile" and other important file system files that cover the first 16 records [1]. During the initial set-up of the file system, a portion of the volume is reserved as "MFT zone," which can be 12.5% of the volume's total size [8].

As new files and directories are created, the MFT file will increase in size and create new record entries. When a file is deleted, the data of its entry in the MFT is wiped and reused for other files. Thus, the MFT file is designed to only increase in size. The file can and often will fragment as its size increases. The size of MFT entries is defined in the boot sector. The most commonly used size is 1024 Bytes for each entry (2 sectors of 512 each), with each entry comprising a header and an attribute area, as shown in Figure 9.2. Each entry is given a sequential number, starting with 0 for the $MFT entry (first entry in the file). Its header consists of 15 fields, with the rest of the entry does not have a pre-defined structure and is primarily used to store various attributes. The end of an MFT entry is identified by an ending mark, with a value of 0X FFFF FFFF.

The MFT header area stores data about the entry itself and has a fixed size and structure, spanning the first 56 bytes of the MFT entry, as shown in Figure 9.3. The first field of the header is a signature that indicates what the MFT entry is referencing. For example, a file is assigned the following signature 0x46494C45. Further entry information given by fields in the header includes "Base FILE record" that indicates if the entry is the base entry of the file (field takes "0" value) or an extended entry (field points to the base entry), "Allocated size of the File record" that indicates whether the file has allocated space and how much that space is.

The attribute area may be fully utilized to store information, although that is not necessary. The attribute area generally stores data structures for the various data types, such as file name, file or directory metadata, and the file content [6]. An important observation about the NTFS is that, unlike other file systems that read and write file content, it instead accesses attributes for files, one of which is the content of a file.

Key elements of MFT entry:

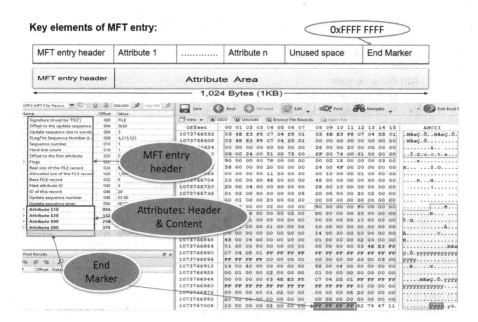

FIGURE 9.2
Structure of MFT entries.

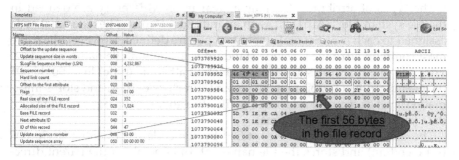

FIGURE 9.3
MFT header example.

Each attribute is separated into a header area and content area [5]. The attribute header area stores several descriptive fields about the content of the attribute, such as the attribute type and length (4 bytes each), attribute name, flags that determine if the attribute is encrypted, compressed, or sparse, as shown in Figure 9.4. The attribute content area stores the actual content of the attribute, which can be the file name, metadata, and content and thus can span several MBs or even GBs. Because the size of the attribute contents may be too large for the MFT file, the content of attributes is separated into two types, "resident data" if the content is included in the MFT file and

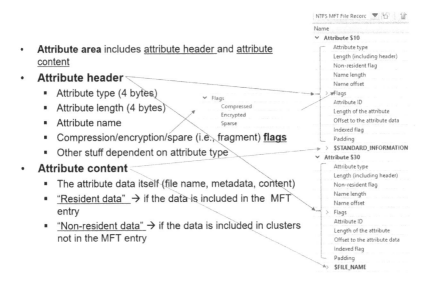

- **Attribute area** includes <u>attribute header</u> and <u>attribute content</u>
- **Attribute header**
  - Attribute type (4 bytes)
  - Attribute length (4 bytes)
  - Attribute name
  - Compression/encryption/spare (i.e., fragment) **flags**
  - Other stuff dependent on attribute type
- **Attribute content**
  - The attribute data itself (file name, metadata, content)
  - "Resident data" → if the data is included in the MFT entry
  - "Non-resident data" → if the data is included in clusters not in the MFT entry

FIGURE 9.4
Attribute area, including header and content.

"non-resident data" if the content is stored elsewhere in the file system and the attribute content stores the cluster location of the actual content.

The types of attributes that may appear in an MFT entry are defined in the file system file "$AttrDef," with an individual unique hexadecimal code assigned to each [3,5]. Several attribute types are defined. However, the three most common attributes include standard information, file name, and data. The code identifies the standard information attribute (type) 16 or $0x10 00$ (little-endian), can be found in all file and directory entries, and stores information such as security information, file/directory ownership, and temporal data (creation/modification time).

The file name attribute is identified by the code (type) 48 or $0x30 00$ (little-endian) and stores the file name, information about the parent directory, and temporal data similar to but less reliable than the standard information attribute. The code identifies the data attribute (type) 128 or $0x80 00$ (little-endian), and its content stores either the contents of a file if it fits in the MFT or data runs, which are collections of clusters in the file that are determined by the first cluster and the length of the run (in clusters). A single file may have multiple data attributes, also known as alternate data streams [12].

Other important attributes of the MFT entries store positional information about the clusters that are used by the file system [14]. The logical cluster number is used to address each cluster in the file system relative to the first cluster in the volume. In contrast, the virtual cluster number is a number given to clusters that belong in a non-resident stream, with the first cluster of each stream given the value "0." NTFS utilizes a dedicated file called

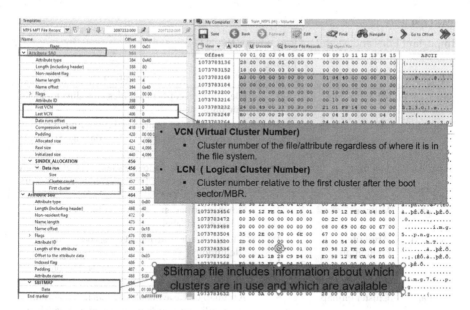

FIGURE 9.5
Example of logical cluster number and virtual cluster number using active disk editor.

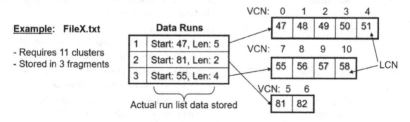

FIGURE 9.6
Fragmented file data run example.

"$Bitmap" to manage the clusters in the volume, where the status of a cluster, either being allocated or unallocated, is stored, as shown in Figure 9.5.

The content of an entry is stored in the "$Data" and can either be found in the MFT file (resident) or they can be stored in multiple areas (fragments) of the file system (non-resident), called data/cluster runs [10]. To reconstruct a file, a data run needs to define the start cluster of the data run (logical cluster number), the length of the data run, and the sequence number if the file is made up of more than one data run. If these runs cannot be stored in a single MFT entry, more entries are assigned to the file. An example can be seen in Figure 9.6.

MFT maintains several reserved entries that log important information that the file system utilizes, such as files that store file system administrative data and metadata for those files. Such files include the $MFT, which is the file that

TABLE 9.2

Important Files and Metadata Records for the NTFS

File Name	File System	Record Position	Description
$MFT	MFT	0	File storing records for all files (and directories) in the file system
$MFTMirr	MFT2	1	Copy of the MFT maintained for redundancy and in case the original MFT file is damaged. Used to recover the MFT
$LogFile	Log file	2	Transactions related to metadata manipulation are stored here in case of a catastrophic event occurs, and recovery is necessary
$ volume	Volume	3	Stores information about the volume such as version and label
$AttrDed	Attribute definition	4	A file that stores attribute types, codes, and definitions
$	Root filename index	5	Represents the root folder of the NTFS volume
$Bitmap	Cluster bitmap	6	A bitmap of the available clusters in the volume that indicates which are allocated and which are free
$Boot	Boot sector	7	File stores code for the bootstrap process of the NTFS file system
$BadClus	Bad clusters	8	Maintains the clusters that have been determined to be damaged and unusable
$Secure	Security file	9	Holds security descriptions for the file system such as the access control list (ACL) for all files and directories in the volume
$Upcase	Upcase table	10	Used to convert all lowercase characters to uppercase Unicode for the volume
$Extend	NTFS extension	11	File with optional NTFS extensions, such as quotas and object identifiers
		7 12–15	Reserved records for future implementations

stores the MFT table. This $Boot file stores the boot code that launches the OS, the $Bitmap file that stores a bitmap of allocated and unallocated clusters in the file system, and the $BadClus file that keeps a map of the identified bad clusters. Table 9.2 lists the vital file system files of the NTFS file system [11].

9.4 NTFS Analytical Implications

In general, the process of recovering data from a hard drive may become necessary during a digital forensic investigation launched for a criminal investigation, for disaster recovery due to corrupted or damaged hard drives, and e-discovery in the context of civil litigation. Such investigations seek to detect hidden or damaged data in hard drives that may later be used as supporting evidence in legal proceedings. Data recovery, in general, can be categorized as either hardware-based recovery that is primarily used to recover data from physically damaged hard drives or software-based. After the hardware-based recovery has been applied, interpret recovered data from either corrupted or physically damaged drives [9].

NTFS utilizes a B-tree structure to index files and directories, with B-tree being a tree-like data structure that supports more than two children in each node and stores sorted data. The allocation status of clusters is stored in the bitmap file, which can be used to identify hidden data that are stored in clusters that appear to be unallocated. In addition, as with FAT deleting a file will not result in immediate deletion of its contents unless the MFT record is overwritten and the file clusters are not deleted in NTFS. First, the root directory can be detected in an MFT entry to analyse a B-tree indexed NTFS volume. The $INDEX_ROOT and $INDEX_ALLOCATION attributes can be processed to identify index records that correspond to nodes of the tree.

9.5 Analysis and Presentation of NTFS Partition

This section explains how to examine an NTFS partition using the Disk Editor and WinHex on Windows 10. Moreover, it explores and generates reports for this parathion using the Autopsy tool for illustrating its forensics analysis.

9.5.1 Disk Editor for NTFS Analysis

The NTFS file system has two important data structures: NTFS Boot Sector and NTFS MFT File Record. Specifically, the MFT is the critical structure of NTFS (a database), where all information about the files and directories is stored. To view the NTFS templates and the elements of the NTFS Boot Sector, open the Disk editor tool and go to "File," select "Partition (H:)," "open," then select "NTFS Boot Sector" from the Templates dropdown. The NTFS Boot Sector can be seen in Figure 9.7. Although the NTFS and FAT boot sectors are similar, there are slight differences in the elements, like the BIOS Parameter Block. Select "NTFS MFT File Record" in the Templates dropdown to view the MFT File Record, as seen in Figure 9.8.

9.5.2 WinHex Tool for NTFS Analysis

The process is similar to the one used for FAT analysis. Start WinHex with administrator privileges (right-click, "Run as Administrator"). Go

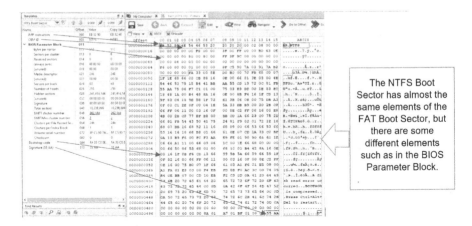

The NTFS Boot Sector has almost the same elements of the FAT Boot Sector, but there are some different elements such as in the BIOS Parameter Block.

FIGURE 9.7
NTFS Boot Sector values of TRAIN_NTFS (H:).

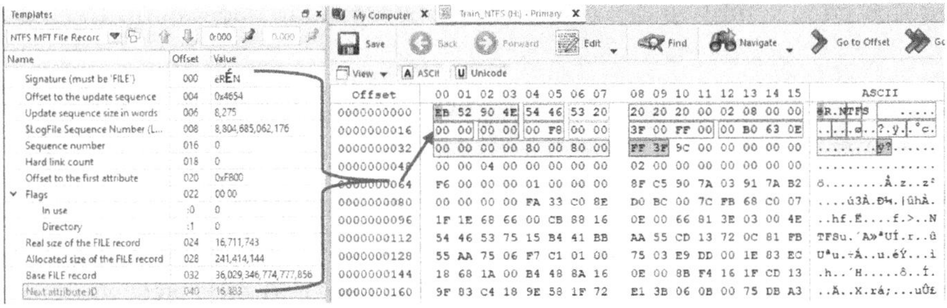

FIGURE 9.8
NTFS MFT of TRAIN_NTFS (H:).

to "Tools," "Open Disk" and select "HD0: VMware, VMware Virtual S (120 GB, SAS)." Then, select "Partition 8 (H:)" and then "View," "Template Manager." In the template manager, to view the boot sector, select "Boot Sector NTFS" and click "Apply!." To view the directory entry, in the template manager, select "NTFS Directory Entry" and click "Apply!," as depicted in Figure 9.9.

9.5.3 The Autopsy Tool for FAT and NTFS Analysis

The autopsy tool is a digital forensics tool that can be used to analyse hard disks and their partitions. Before any analysis can occur, you need to validate the integrity of the images you will examine for evidence (hash functions, comparison of hash values). First, delete some files from partition (I:) (FAT32). Next, launch Autopsy and create a new case for the analysis of the partition. Ensure that you have selected "local disk" from the "Add Data Source"

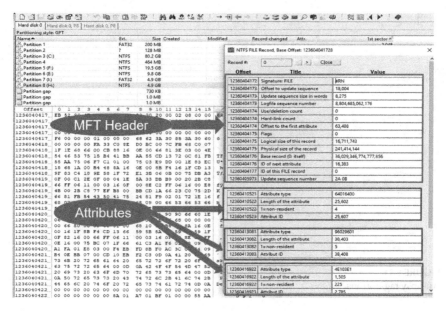

FIGURE 9.9
Analysis of MFT header and attributes using WinHex.

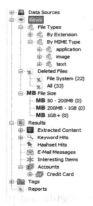

- **Views** provide information about files and their types, their sizes as well as deleted files.
- **Results** provide information about modules used, such as finding information about keywords, emails and accounts.
- **Tags** provide information about bookmarks that you could do further analysis.
- **Reports** include the presentation step of digital forensics to generate reports about your investigation.

FIGURE 9.10
Autopsy left bar.

window. After the partition is loaded in Autopsy, you should see a collection of key elements in the left bar, as seen in Figure 9.10.

These key elements are "Views" that provide information about file types and size, including deleted files, "Results" that provide results of modules used to search information about keywords, emails, and accounts, "Tags" where information about elements that you have tagged is provided and

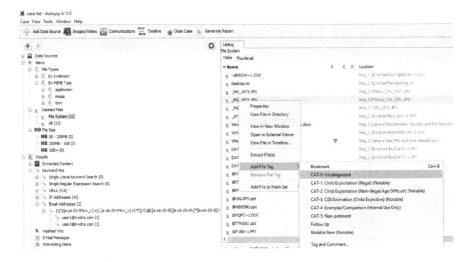

FIGURE 9.11
Autopsy tag options.

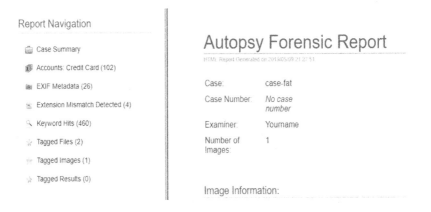

FIGURE 9.12
Autopsy HTML report.

"Reports" where reports are generated and used for the presentation phase of a digital forensic investigation.

Regarding the "Tags" elements, when investigators identify an interesting element, they can right-click and select one of the pre-defined tag options, as seen in Figure 9.11. After tagging it, the file can be found under the "Tags" element in the left bar, and it can be included in any generated report.

After the analysis phase is over, select "Generate Report" from the top bar to generate a report. Supported report extensions include HTML, Excel, and text. Next, select "ALL Results" and then "Finish." An example of an HTML report on the FAT32 partition can be seen in Figure 9.12. Clicking on any element in the "Report Navigator" will present more information related to the element (for example, tagged files).

9.6 Conclusion

In this chapter, we discussed the elements of the NTFS file system, including the reserved files and directories. Additionally, we explained the structures that NTFS uses to store data and presented the digital forensic analysis and implications of NTFS volumes. Furthermore, the NTFS was inspected using the Disk Editor, Winhex, and Autospys tools to find evidence and produce forensics reports. Next, the FAT and NTFS Recovery will be investigated.

References

1. Chapter 2: Systems, disks, and media. In: Dave Kleiman, Kevin Cardwell, Timothy Clinton, Michael Cross, Michael Gregg, Jesse Varsalone, and Craig Wright (editors), *The Official CHFI Study Guide (Exam 312–49)*, pp. 61–131. Syngress, Rockland, 2007.
2. EC-Council. *Computer Forensics: Investigating File and Operating Systems, Wireless Networks, and Storage (CHFI)*. Computer Hacking Forensic Investigator. Cengage Learning, Boston, MA, 2016.
3. Steve Anson, Steve Bunting, Ryan Johnson, and Scott Pearson. *Mastering Windows Network Forensics and Investigation*. ITPro collection. Wiley, Hoboken, NJ, 2012.
4. LLC Books and S. Wikipedia. Windows Disk File Systems: File Allocation Table, Ntfs, High Performance File System, Winfs, Robocopy, Virtual Folder, Ntfs Junction Point. General Books LLC, 2010.
5. Steve Bunting and William Wei. *EnCase Computer Forensics: The Official EnCE: EnCase?Certified Examiner Study Guide*. Wiley, Hoboken, NJ, 2006.
6. Brian Carrier. Digital crime scene investigation process. In: *File System Forensic Analysis*, pp. 13–14. Addison-Wesley Professional, Boston, MA, 2005.
7. Xinyu Feng and Sungwoo Park. *Programming Languages and Systems: 13th Asian Symposium, APLAS 2015*, Pohang, South Korea, November 30 to December 2, 2015. Lecture Notes in Computer Science. Springer International Publishing, 2015.
8. Nihad Ahmad Hassan and Rami Hijazi. Chapter 4: Data hiding under windowsÂ R os file structure. In: Nihad Ahmad Hassan and Rami Hijazi (editors), *Data Hiding Techniques in Windows OS*, pp. 97–132. Syngress, Boston, MA, 2017.
9. Nora Haynes. *Cyber Crime*. EDTECH, 2018.
10. Xiaodong Lin. *Introductory Computer Forensics: A Hands-on Practical Approach*. Springer International Publishing, Wilfrid Laurier University Waterloo, ON, Canada 2018.

11. Amelia Phillips, Bill Nelson, and Christopher Steuart. *Guide to Computer Forensics and Investigations.* Cengage Learning, Boston, MA, 2014.
12. Michael Raggo and Chet Hosmer. Chapter 7: Operating system data hiding. In: Michael Raggo and Chet Hosmer, editors, *Data Hiding*, pp. 133–166. Syngress, Boston, MA 2013.
13. Marcus Rogers and Kathryn Seigfried-Spellar. *Digital Forensics and Cyber Crime: 4th International Conference, ICDF2C 2012,* Lafayette, IN, USA, October 25–26, 2012, Revised Selected Papers. Lecture Notes of the Institute for Computer Sciences, Social Informatics and Telecommunications Engineering. Springer Berlin Heidelberg, 2013.
14. David A. Solomon, Mark E. Russinovich, Alex Ionescu. *Windows Internals.* Developer Reference. Pearson Education, London, 2009.
15. Thomas Sterling, Matthew Anderson, and Maciej Brodowicz. Chapter 18: File systems. In: Thomas Sterling, Matthew Anderson, and Maciej Brodowicz (editors), *High Performance Computing,* pp. 549–578. Morgan Kaufmann, Boston, MA, 2018.

10

FAT and NTFS Recovery

10.1 Introduction

In the previous two chapters, we discussed the two most popular file systems, FAT and NTFS, commonly found in computer systems. We observed how these file systems store, retrieve, and manage files and discussed their differences. In this chapter, we will be focusing on the process of recovering deleted files from either of the two file systems. Furthermore, we will discuss what happens to files when they are moved to the recycling bin by the user. Then, we use various recovery tools and demonstrate how they can be employed in some realistic scenarios.

The main objectives of this chapter are as follows:

- Understand the NTFS and FAT file system recovery
- Explain recycle bin operations
- Apply recovery tools, including Foremost, Scalpel, and Bulk Extractor, to file systems and images

10.2 FAT and NTFS File Recovery

Deleting a file does not necessarily equate to the loss of the content of a file. Most modern file systems do not erase the contents of a file when it is deleted, but in fact, mark the file's entry in the file system tables (directory table in FAT or MFT in NTFS) as being available for reuse [5,6]. For an NTFS partition, a flag at offset 22 is set to "0x0000," which means the entry is "not in use," and the file is considered deleted. For a FAT partition, the "0xE5" character is attached to the beginning of the file's name, thus indicating that the directory entry of this file can be reused. If the entries have not been overwritten with the metadata of new files and their content clusters are intact, files can be quickly recovered.

DOI: 10.1201/9781003278962-10

10.2.1 Deleting and Recovering Files in FAT File System

Regardless of the file system in use, the MBR is present at the beginning of the hard drive and will contain information about the logical partitioning of the medium. The MBR loads and forwards control to the boot sector, which is the first sector of a partitioned space, and from there, the OS is loaded, and the file system is active [20]. In the FAT table, there can be four types of entries, "allocated" which represents an active cluster of a file and is accompanied by a link to the next cluster allocated to the file, "unallocated" which is a free cluster represented by the value "0x0000," "EOF" which is a special character sequence indicating the end of a file in a cluster chain (for FAT16 EOF is "0Xffff") and "bad cluster" which represents a damaged cluster that cannot be used by the file system (for FAT16 represented by "0xfff7").

When a file is deleted, the special character "0XE5" is attached at the beginning of the file's name, with the rest of the directory entry remaining unaffected, as shown in Figure 10.1 [10,11]. However, the FAT table entry that the directory entry for the deleted file points to as the first cluster is affected. Specifically, the FAT table entry no longer points to the next cluster or EOF for the file and instead is set to "0," indicating that the FAT entry is available for reuse. This could hinder the recovery of a file, although investigators often employ an effective strategy if the file is not fragmented. By identifying the first cluster of the file, and the size of the file, knowing the size of a cluster, an investigator can retrieve the next "X" clusters as part of the file, with "X"

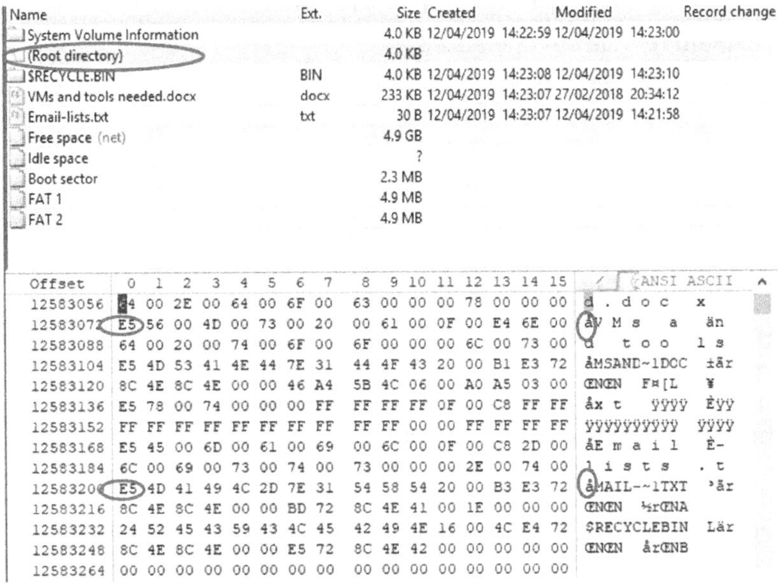

FIGURE 10.1
Example of finding E5 of deleted files in FAT.

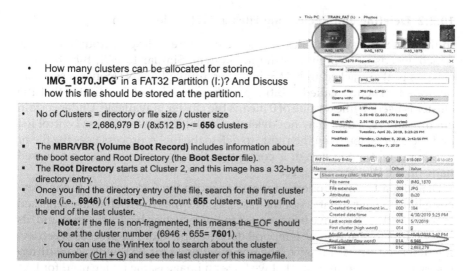

- How many clusters can be allocated for storing
  'IMG_1870.JPG' in a FAT32 Partition (I:)? And Discuss
  how this file should be stored at the partition.

- No of Clusters = directory or file size / cluster size
  = 2,686,979 B / (8x512 B) ~= **656** clusters

- The **MBR/VBR (Volume Boot Record)** includes information about
  the boot sector and Root Directory (the **Boot Sector** file).
- The **Root Directory** starts at Cluster 2, and this image has a 32-byte
  directory entry.
- Once you find the directory entry of the file, search for the first cluster
  value (i.e., **6946**) (**1 cluster**), then count **655** clusters, until you find
  the end of the last cluster.
  - **Note:** if the file is non-fragmented, this means the EOF should
    be at the cluster number (6946 + 655= **7601**).
  - You can use the WinHex tool to search about the cluster
    number (<u>Ctrl + G</u>) and see the last cluster of this image/file.

FIGURE 10.2
Example of FAT cluster allocation.

being equal to the size of the file divided by the size of a cluster (minus 1 for
the initial cluster).

For example, assume we are trying to recover a deleted image file titled
"IMG_1879.jpg" in a FAT32 partition and let us assume its size is equal to
2,686,979 Bytes, as depicted in Figure 10.2. To determine the number of clusters necessary to store this file, we would divide the size of the file by the total
size of a single cluster that is equal to 8 sectors, and each sector is equal to
512 Bytes. Thus, the total number of sectors would be 2,686,979/(8∗512) equal
to 656 clusters. By reviewing the directory table and partition, we can locate
the directory entry of the deleted file and identify its initial cluster. Then,
assuming the file is sequential, we can load the next 655 clusters as the file's
content. Tools such as WinHex [3] can be used to search the cluster numbers
for this process.

To recover a file, two main strategies may be employed [8,19]. If the files or
directories that we are attempting to recover were not fragmented, their clusters are still intact and have not been overwritten or wiped in any way, recovery tools can be used to restore the original directory entries, and the files
can become accessible again. This process is known as automatic recovery.

Suppose, on the other hand, the file or directory was not sequentially stored
(fragmented). In that case, the recovery process relies on either detecting the
first cluster and extracting the rest sequential clusters, as was described in
the previous paragraph, or extracting clusters after the initial until the EOF
sequence is detected. Then, the collected data is assembled in a single file,
proper file types are selected, and the file is accessed. This is known as manual recovery.

10.2.2 Deleting and Recovering Files in NTFS File System

In the NTFS file system, files and directories are indexed through a B-tree, which speeds up searches in the file system. Similar to FAT, the status of clusters is also tracked in NTFS; however, unlike FAT, NTFS has a separate dedicated structure called "$Bitmap," where a value of 1 indicates that the cluster is used, while a value of 0 means the cluster is available. Under certain circumstances, a user may not delete a file [16] if it has an access control list and the user is not given the rights to delete the file. A process uses the file or another user, the file system is corrupt, or the file name includes a reserved name in the Win32 namespace.

Deleting a file in Windows appears to move the file's content to the recycling bin; however, that is not the case [21]. The files themselves are renamed to indicate that they have been moved to the recycling bin by the user and, its parent directory and some temporal data are updated. The process of retrieving a deleted file is similar to the one mentioned for FAT, with some differences. For example, assume that we have an image (same image used for FAT example) stored in a partition (in the Root directory). By accessing the VBR, information about the root directory can be identified, and thus MFT entries can be accessed to find all reserved files (starting with "$").

The exact location of the Root directory (in clusters) is not pre-defined, and to find it, divide the number of sectors until the root directory is reached by 8 (8 sectors=1 cluster)- an example of allocating NTFS Cluster shown in Figure 10.3. Next, after accessing the MFT file and the MFT entry for the file we are attempting to retrieve has been identified, the "$AttrDef" is used to detect the location of the first cluster of the file. Then, by accessing the "$Bitmap" file, we begin at the first cluster of the file and access the next "X" clusters, where "X" is equal to (number_of_clusters_of_file -1). Similar to the FAT case, WinHex can be used to search for the clusters.

10.3 Recycle Bin and Forensics Insights

Although deleting a file in Windows results in the file being moved to the recycle bin, that is not always the case. Under certain circumstances, files may be immediately deleted, thus not moved to the bin [15,17]. For example, deleting files from external media such as USB sticks will be erased. In addition, cases where files are deleted from the DOS prompt, network shares, system deleted files, or too big files to fit in the recycle bin all bypass the recycle bin. As the recycle bin has a set size, it will reach its maximum size at some point. At that time, older files will be wiped to make room for newer files that are moved to the recycle bin, with the deletion methodology functioning in a first-in-first-out mode.

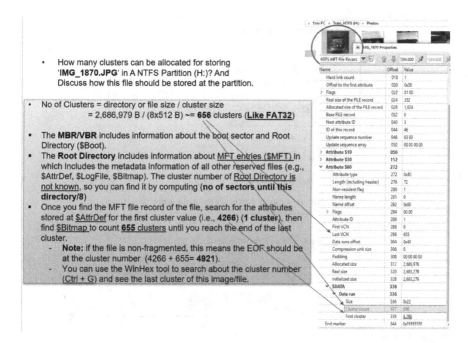

- How many clusters can be allocated for storing 'IMG_1870.JPG' in A NTFS Partition (H:)? And Discuss how this file should be stored at the partition.

- No of Clusters = directory or file size / cluster size
  = 2,686,979 B / (8x512 B) ~= **656** clusters (**Like FAT32**)

- The **MBR/VBR** includes information about the boot sector and Root Directory ($Boot).
- The **Root Directory** includes information about MFT entries ($MFT) in which includes the metadata information of all other reserved files (e.g., $AttrDef, $LogFile, $Bitmap). The cluster number of Root Directory is not known, so you can find it by computing (**no of sectors until this directory/8**)
- Once you find the MFT file record of the file, search for the attributes stored at $AttrDef for the first cluster value (i.e., **4266**) (**1 cluster**), then find $Bitmap to count **655 clusters** until you reach the end of the last cluster.
  - **Note:** if the file is non-fragmented, this means the EOF should be at the cluster number (4266 + 655= **4921**).
  - You can use the WinHex tool to search about the cluster number (Ctrl + G) and see the last cluster of this image/file.

FIGURE 10.3
Example of allocating NTFS cluster.

The recycle bin has certain properties with default settings that affect its behaviour. These settings can be customized by the user and include the size of the bin, which is 10% of the drive size by default. There is an option to not send files to the recycle bin and immediately remove them, with the default setting being to send files to the bin first and a confirmation dialogue prior to filing deletion, with the default setting being to display the message. An example of these settings can be seen in Figure 10.4.

The recycle bin creates a hidden system folder, where deleted files are moved, with the hidden folder named "RECYCLER" in XP systems [12] and "$RECYCLE.BIN" in newer NTFS and FAT32 systems starting with Windows 7 [7]. Moving files to these hidden directories result in changes to some of the file's metadata. For instance, the parent directory information of the file is changed to represent the new parent directory. Other values such as creation/modification/access times, file size, and first content cluster are retained. It is important to mention that each partition has its own "$RECYCLE.BIN," as shown in Figure 10.5. Thus, if files are deleted in different partitions that may be mounted to the main OS partition, then these files will be found in the recycle bin of the partition where it was deleted.

In order to restore deleted files from the recycle bin, in Windows XP and earlier versions, certain metadata information about the files, such as deletion

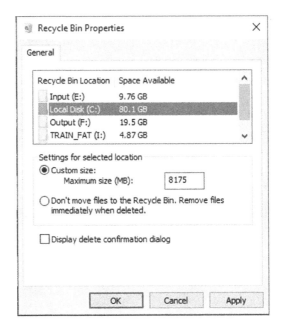

FIGURE 10.4
Example of recycle bin settings.

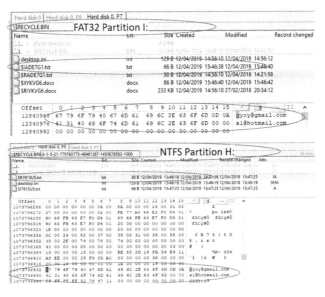

FIGURE 10.5
Example of $RECYCLE.BIN in FAT and NTFS partitions.

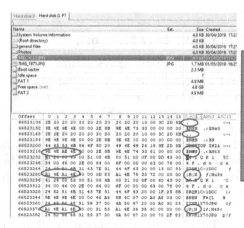

- It starts with ".")Root Directory, then ".." (Current Director ($RECYCLE.BIN))
- Any deleted file in FAT32 partitions should have **4 bytes** as **a moved maker** the ASCII (-N-N) and (- 4E - 4E) in hexadecimal, where "-" depends on the file types and directory types.

FIGURE 10.6
Example of analysing recycle bin in FAT.

date/time, drive number of origin, the original path of file and size, are stored in special files, INFO and INFO2 [14]. Another important fact about the recycle bin in an NTFS partition is that each user that accesses a computer has an individual hidden subdirectory inside "$RECYCLE.BIN." Thus, when a user deletes a file, it is moved to their personal recycle bin subdirectory, named based on the user's security identifier, assigned to the user by Windows OS [13,14].

An example of the recycle bin in FAT32 would include references to the current directory "." and the enclosing directory "..," which is the root directory. Additionally, each file that is moved to the recycle bin should have 4 bytes "0x -4E -4E" in hexadecimal and "- N - N" in ASCII to mark them, with the "-" value being dependent on the file and directory types. In NTFS, the "$RECYCLE.BIN" is stored in a 1 KB (2 sectors) MFT record, and it includes the marker "FILE" at the beginning of its entry. The first entry in the "$RECYCLE.BIN" is for the bin itself, and each file that is moved to the bin has to end with the marker "0XFFFF FFFF" (Figure 10.6).

When a file is deleted in NTFS, two files are created and named pseudo-randomly, one starting with "$I" with the original file's metadata and one starting with "$R" with the file's data itself [14,18]. Deleting a file from a removable media results in the file being wiped immediately, as it circumvents the recycle bin. In Windows 10, the security identifier for the system admin would begin with the security identifier code "S-1-5-21," and thus the recycle bin for that user would similarly start with the same code. The entry for the recycle bin of that user can be found in the registry editor, which can be accessed by typing "regedit" at the Windows search bar.

10.4 Mounting Partitions Using SMB over Network

Mounting is an access point from one partition or operating system to another, enabling sharing files and directories and recovering them. The Windows 10 and Kali Linux virtual machines can be used for mounting partitions of windows to explore them at Kali using SMB protocol over a network. To share a partition on the Windows 10 VM, select a partition to share, for example, Partition C. Right-click on the partition and navigate to "Properties," "Sharing," "Advanced Sharing," select "Share this folder" and then go to "Permissions" and check the "Allow" box for "Read" for everyone. This will make the partition available through the network in a read-only mode.

On the Kali VM, navigate to "Files," "Other Location" and in the "Connect to Server" insert the following address: "smb://192.168.159.151," where "smb" [9] is a network protocol primarily used for providing shared access to files, printers and other such network entities in a network, followed by the IP address of the machine that shares the drive/directory. A login window will request login credentials, the same as the ones used to login to Windows 10 VM (user account: username: "windows10," password: "admin"). This will grant access to the contents of Partition C on the Kali VM, as seen in Figure 10.7.

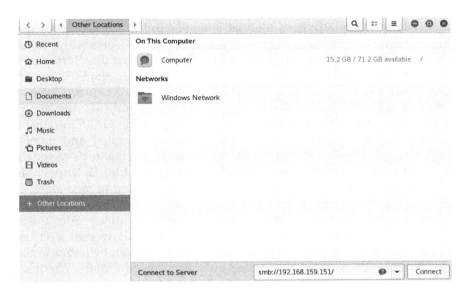

FIGURE 10.7
Example of mounting drive in Kali Linux from Windows 10 using SMB.

10.5 File Recovery and Data Carving Tools for File Systems

Numerous tools exist in which forensics analysts and investigators can use to recover deleted data or traverse a hard drive and carve otherwise unrecoverable files after mounting the files and directories, as explained below.

10.5.1 Foremost Tool

Foremost tool [1] is an open-source forensic recovery tool included in Kali Linux distributions. It utilizes header, footer, and internal data structures to retrieve and carve data from a forensic disk image or a physical drive. The key difference between recovering and carving files is that recovering files utilizes the file system to acquire enough information to locate and retrieve them while carving works on raw data and uses information about the file type such as file signatures to identify and retrieve files. Foremost is a terminal-based and file system independent tool that works on portions of the hard drive, searching for the target file type through carving and file signatures.

To review the functionality of the Foremost tool, first launch the Windows 10 and Kali Linux VMs. Foremost comes pre-installed in Kali Linux so that we will navigate to "Applications," then select "11-Forensics" and select Foremost. This launches Foremost in a new terminal interface, with the available options given in Figure 10.8.

To use Foremost, the syntax is as follows "foremost -I Path_to_forensic_ image -o Path_to_output," with "-i" defining the input of the program and "-o" the output directory, which needs to be empty. After Foremost has completed the carving process, its output can be found in the output directory that was given after the "-o" attribute, organized into sub-directories based on file types, along with an "audit.txt" file that details the findings of the program, as seen in Figure 10.9. Operating Foremost to carve files takes some considerable time.

10.5.2 Scalpel Tool

Scalpel [2] is an open-source program, based initially on Foremost and compatible with Linux and Windows machines. Like Foremost, Scalpel utilizes a database of known file headers and footers to identify and carve them from a disk or image. Its functionality includes regular expressions for headers/ footers, asynchronous I/O that enables concurrent operations for faster results, multithreading, and GPU acceleration.

To operate Scalpel, unlike Foremost, the file types that are to be identified in a forensic image and carved need to be specified in the "scalpel.conf" file, which is located at the "/etc/scalpel/" directory, and at least one file type needs to be provided. Specifically, the comments specifying the file types

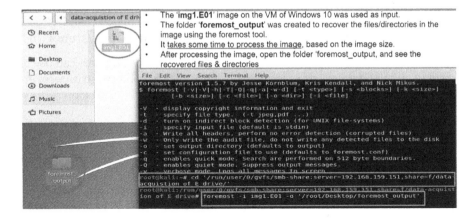

FIGURE 10.8
Foremost example.

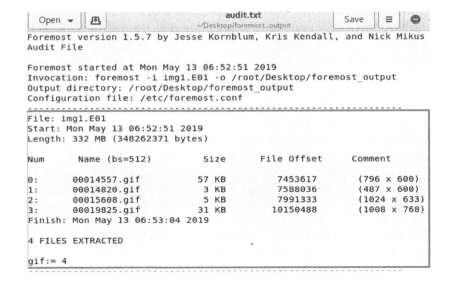

FIGURE 10.9
Foremost audit.txt example.

need to be removed (remove the "#" at the beginning of the line) to have Scalpel retrieve them, as seen in Figure 10.10.

The software can be found in Kali Linux by navigating to "Applications," then "11-Forensics," "Forensics Carving Tools," and then selecting "scalpel" to launch the program in a new terminal window. The syntax for using Scalpel is "scalpel -o Path_to_output_directory Path_to_input_file," and an example of launching Scalpel can be seen in Figure 10.11. Depending on the lines that were uncommented, it may take some time

FIGURE 10.10
Configuration file of Scalpel.

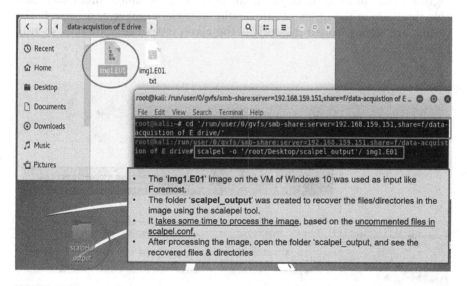

FIGURE 10.11
Recovering image files from forensic image example with Scalpel.

to process an image. The produced output is the same as Foremost is when applied to the same forensic image, with its output grouped into sub-directories based on file type and an "audit.txt" file that describes the program's findings.

10.5.3 Bulk Extractor Tool

Bulk Extractor [4] is a C^{++} tool that can process disk images, files, and directories to extract important features, such as email addresses, URLs, and credit card numbers. It does not interact with file system structures. Instead, it loads entire pages from the drive and extracts information stored in feature files, along with statistical information about the extracted features. Bulk Extractor can be accessed through the terminal or a graphical user interface.

Bulk Extractor is carving software, but unlike Foremost and Scalpel that primarily extract images, video, audio, and compressed files, it can extract a wider range of file types. Data that can be extracted with Bulk Extractor include credit card numbers, emails, URLs, online search history, website information, and social media profiles and information. To access Bulk Extractor through Kali Linux, insert the program's name in the search textbox.

The syntax for running Bulk Extractor is "bulk_extractor -o Path_to_output_directory Path_to_input_image," similar to Scalpel, as shown in Figure 10.12. Running Bulk Extractor on a forensic image takes considerable time due to the range of information it seeks to carve and extract. The results are stored in text files under the given output directory, as seen in Figure 10.13.

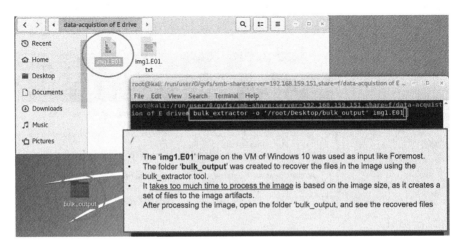

FIGURE 10.12
Recovering image files from forensic image example with bullk_extractor.

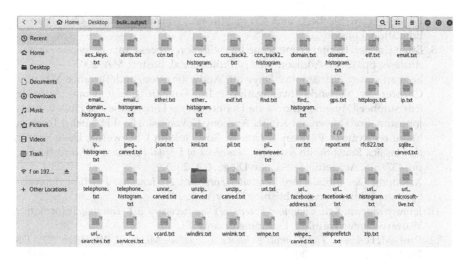

FIGURE 10.13
Bulk extractor output example.

10.6 Conclusion

This chapter discussed the file recovery process for NTFS and FAT partitions. The recycle bin operations and the differences in old and newer Windows OS versions are described, including available options and their role in recovering deleted files. Finally, three commonly used recovery tools, namely Foremost, Scalpel, and Bulk Extractor, were discussed to demonstrate how they recover deleted files and find evidence while recovering files and directories. Next, Linux foundations for digital forensics will be discussed.

References

1. Foremost tool, http://foremost.sourceforge.net/, 2021.
2. Scalpel tool, https://www.kalilinux.in/2021/01/scalpel-recover-permanently-deleted-files-linux.html, 2021.
3. Winhex tool, https://www.x-ways.net/winhex/, 2021.
4. Bulkextractor tool, https://www.kali.org/tools/bulk-extractor/, 2021.
5. EC-Council. *Computer Forensics: Investigating File and Operating Systems, Wireless Networks, and Storage (CHFI).* Computer Hacking Forensic Investigator. Cengage Learning, Boston, MA, 2016.
6. Steve Anson, Steve Bunting, Ryan Johnson, and Scott Pearson. *Mastering Windows Network Forensics and Investigation.* ITPro collection. Wiley, Hoboken, NJ, 2012.

7. Ed Bott, Carl Siechert, and Craig Stinson. *Windows 7 Inside Out*. Pearson Education, London, 2009.

8. Brian Carrier. Digital crime scene investigation process. In: *File System Forensic Analysis*, pp. 13–14. Addison-Wesley Professional, Boston, MA, 2005.

9. Eoghan Casey, Christopher Daywalt, Andy Johnston, and Terrance Maguire. Chapter 9: Network investigations. In: Eoghan Casey, Cory Altheide, Christopher Daywalt, Andrea de Donno., Dario Forte, James O. Holley, Andy Johnston, Ronald van der Knijff., Anthony Kokocinski, Paul H. Luehr, Terrance Maguire, Ryan D. Pittman, Curtis W. Rose, Joseph J. Schwerha, Dave Shaver, and Jessica Reust Smith (editors), *Handbook of Digital Forensics and Investigation*, pp. 437–516. Academic Press, San Diego, CA, 2010.

10. Xiaodong Lin. *Introductory Computer Forensics: A Hands-on Practical Approach*. Springer International Publishing, New York, 2018.

11. Jeremy Martin. *Kali: Computer Forensics Data Recovery 101- Training*. Information Warfare Center Training, Escambia County, FL.

12. Paul McFedries. *The Complete Idiot's Guide to Microsoft Windows XP*. DK Publishing, London, 2001.

13. TJ O'Connor. Chapter 3: Forensic investigations with python. In: *Violent Python*, pp. 81–123. Syngress, Rockland, 2013.

14. Ryan D Pittman and Dave Shaver. Chapter 5: Windows forensic analysis. In: Eoghan Casey, Cory Altheide, Christopher Daywalt, Andrea de Donno., Dario Forte, James O. Holley, Andy Johnston, Ronald van der Knijff., Anthony Kokocinski, Paul H. Luehr (editors), Terrance Maguire, Ryan D Pittman, Curtis W Rose, Joseph J Schwerha, Dave Shaver, and Jessica Reust Smith (editors), *Handbook of Digital Forensics and Investigation*, pp. 209–300. Academic Press, San Diego, CA, 2010.

15. David Pogue. *Windows XP Home Edition: The Missing Manual*. Missing manual. Pogue Press/O'Reilly, California, 2004.

16. Don Poulton. *MCTS 70–680 Cert Guide: Microsoft Windows 7, Configuring*. Cert Guide. Pearson Education, London, 2010.

17. John Sammons. *The Basics of Digital Forensics: The Primer for Getting Started in Digital Forensics*. Elsevier, Amsterdam, 2012.

18. Oleg Skulkin and Scar de Courcier. *Windows Forensics Cookbook*. Packt Publishing, Birmingham, 2017.

19. Richard Smith. *Elementary Information Security*. Jones & Bartlett Learning, Birmingham , 2019.

20. Thomas Sterling, Matthew Anderson, and Maciej Brodowicz. Chapter 18: File systems. In: Thomas Sterling, Matthew Anderson, and Maciej Brodowicz (editors), *High Performance Computing*, pp. 549–578. Morgan Kaufmann, Boston, MA, 2018.

21. Ahmed Bahjat and Jim Jones. Deleted file fragment dating by analysis of allocated neighbors. *Digital Investigation*, 28:S60–S67, 2019.

11

Basic Linux for Forensics

11.1 Introduction

In previous chapters, we have employed two well-known Linux operating system (OS) distributions, Ubuntu and Kali Linux, to perform forensic tasks, such as validating files by using hash functions and creating forensic images of drives. In this chapter, we focus on the Linux OS and its components. Furthermore, we will discuss how Linux operates with different partition styles and file systems.

The main objectives of this chapter are as follows:

- Learn key features of Linux: kernel, networking, and file systems
- Explain Master Boot Record (MBR) and GUID Partition Table (GPT) styles in Linux systems
- Understand how Linux OS uses different file systems
- Apply the hard disk analysis in Linux using the wxHexEditor and Active Disk Editor tools
- Execute data acquisition in Linux using the dd and dcfldd commands

11.2 Overview of Linux Operating System

Linux is an open-source OS that started as a clone of Unix in 1991 [15]. It was developed from scratch by Linus Torvalds, who was assisted by a globally distributed team of hackers. Unix OS was developed in 1969 by a group of AT&T employees at Bell Labs as multitasking and multi-user OS, intended as a platform for software development [25]. An OS can be considered a "middle-man" between a computer's hardware and its applications, providing services to the applications, and translating their requests to instructions that can be executed on the underlying hardware.

Today, Linux OS can be found virtually everywhere, from PCs and laptops to servers, smartphones, and embedded devices of the Internet of Things (IoT) [23]. It is open source and, unlike other OSs, provides software tools at no cost. Linux is a complete OS robust and stable, as a program crash is unlikely to disrupt the OS operations [20]. Linux is reliable, as Linux servers often remain up and running for hundreds of days, compared to Windows systems that require regular reboots and are extremely powerful. It provides a complete development environment, including libraries for low-level OS interactions and other services, compilers, and debuggers [29,30].

Furthermore, Linux OS provides excellent networking facilities that are easily configurable by the users. It is an ideal environment to run servers such as web, FTP, and email servers. It is easily upgradeable and true multi-tasking and multi-user OS that supports multiple processors for greater compatibility. In addition, apart from the free software that comes with Linux, a variety of commercial software is also available. Finally, as Linux OS is open source, its source code is freely available to anyone. The popularity of Linux OS is evident by the number of available variant distributions, with each distribution often incorporating the Linux core (called Kernel) and different applications and libraries and thus providing slightly different services to its users. Some popular distributions include Ubuntu [8], Debian [2], Arch [5], Fedora [4] and CentOS [6].

The Linux system is, by design, reliable and robust. It excels in many areas, ranging from end-user concerns such as stability, speed, and ease of use to serious concerns such as development and networking. This is because it has been refined over the years by hundreds of contributors from around the globe due to its open-source nature. The Linux system can be separated into three main components: the Kernel, networking, and file system.

11.3 Linux Kernel

The Linux Kernel can be considered as the central system of Linux OS [16]. It is code that lies at the core of an OS, and it controls everything in a computer, providing resources and managing every program executed in a Linux machine. The Kernel is loaded when the machine boots and remains in the memory (RAM) as long as the computer is on. As depicted in Figure 11.1, it connects the hardware to software applications, allowing them to use hardware services such as memory space, I/O devices, and TCP/IP networking during their execution. The Kernel is also responsible for managing processes and threads, switching between them to simulate the concurrent performance of programs.

The Linux kernel is designed modularly, making it highly extensible and allowing actual OS code to be very small, enabling it to be swiftly loaded into

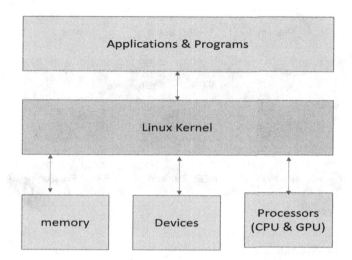

FIGURE 11.1
Linux kernel with applications, memory, devices, and processors.

memory without requiring extensive space. Linux is, in a sense, a product and an integral part of the Internet, with many servers, routers, and other network entities running some variant of Linux OS [26]. It supports a wide range of networking protocols, with the most important ones being the TCP/IP protocol suite that makes up what is known as the Internet. It should be mentioned that the Kernel manages networking operations.

11.4 Linux File System

A file system is an abstract system that enables the creation, deletion, modification, and management of files organized under directories. All OSs utilize a file system to manage their files, and Linux is no exception. In Linux, files are classified using the File System Hierarchy standard, which defines the directory structure and contents of Linux OS and its variants.

Current Linux systems employ what is known as a virtual file system (VFS) [22], which functions as an additional layer of abstraction that operates over other, well-established file systems, allowing users and applications to access files from the underline file systems in a uniform manner, as shown in Figure 11.2. OSs like Linux or Windows can mount multiple storage drives or partitions, each with a different underline file system. Systems would be overcomplicated if threads accessing files on mounted drives had to use various APIs for every distinct underline file system [21]. This is the problem that virtual file systems solve by playing the middle-man role, as they provide POSIX API to processes and forward their requests to low-level file system

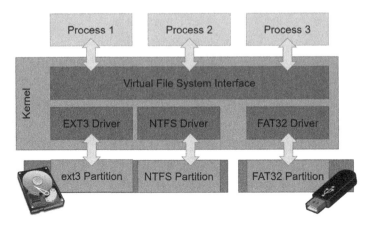

FIGURE 11.2
Flowchart of a virtual file system.

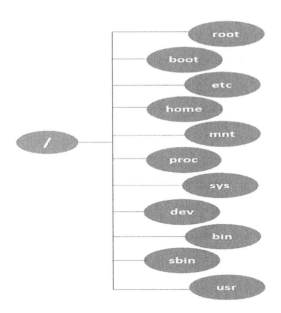

FIGURE 11.3
Linux file system structure Linux systems allow hard disk drives to be partitioned into two or more separate physical entities, each with its file system.

drivers. The virtual file system is kernel software, and this method is followed by both Linux and Windows OS.

Under the File System Hierarchy standard, all directories, including those stored on separate storage media but have been mounted, are organized under the root directory "/" and appear in a tree-like structure, as shown in Figure 11.3. In the tree-like structure, "/home" is where a user's directory is located, the superusers home directory is in "/root," "/boot" holds the Linus OS kernel image, system configuration files are under "/etc," devices are mounted under "/mnt," "/bin" and "/sbin" hold executables and "/lib" the libraries that can be used by software to interact with services provided by the OS.

11.4.1 Linux Hard Drives and Styles

The Linux system is compatible with MBR and GPT disk architectures, like Windows OS. Unlike Windows, naming conventions for drives in Linux do not permit the use of letters like "C" or "D"; instead, strings like "/dev/sdc" or "/dev/sdd" are used. The "dev" in a drive's name stands for "drive," referred to as "block storage devices" in Linux because the system interacts with the storage medium in fixed block sizes [28]. The "sd" part of the string that precedes the letter of the drive refers to "SCSI mass-storage Driver" with "SCSI" being a set of standards for transferring data on a physical level, between computers and peripheral devices (e.g. printer and hard disk). Drives in Linux are considered to be files, and thus the names of drives (e.g. "/dev/sda") are actually the full path that point to the files that represent the drives [18].

In the Linux file systems, file names are limited to 256 characters, and slash ("/") is used for the path of directories and files, as opposed to the backslash ("\") that is used in DOS. Several file types are supported by Linux file systems [15]. Such file types include text, command, data and executable files, directories, special device files, and links. Linux uses link files to preserve disk space, with the two types of link files being hard-linked files that function as a duplicate of a file and soft link files that function as references to other files [27]. Deleting the original file does not affect a hard link file, although altering the original file's content will have the same effect on the hard link. At the same time, it invalidates a soft link file, as it functions as a pointer to the original.

In Ubuntu, information about the existing partitions can be viewed through the use of specific software tools. One such tool is the "gnome-disks" utility [10], which can be accessed through the terminal after acquiring administrative privileges. It displays the volumes of the hard disk and information about them, as shown in Figure 11.4. For the Ubuntu Server 14 VM, the partitioning scheme is MBR, and thus all partitions that are not explicitly created as extended partitions, are by definition, primary partitions, with a maximum of four primary partitions in any disk (or three primary and one extended). The volume includes Partition1 ("/dev/sda1") where MBR is located, an Extended partition (sda2), a Swap partition (sd5), and a File system partition (Ext4).

A similar tool that can be accessed solely from the terminal is "fdisk" [1]. It requires superuser privileges, similarly to "gnome-disks," and it provides extensive information about the connected hard disk devices and their properties. Figure 11.5 lists the total number of heads, sectors, cylinders, and

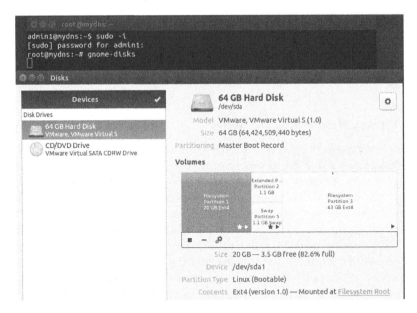

FIGURE 11.4
Example of Gnome-disk utility.

```
root@mydns: ~
admin1@mydns:~$ sudo -i
[sudo] password for admin1:
root@mydns:~# fdisk -l

Disk /dev/sda: 64.4 GB, 64424509440 bytes
255 heads, 63 sectors/track, 7832 cylinders, total 125829120 sectors
Units = sectors of 1 * 512 = 512 bytes
Sector size (logical/physical): 512 bytes / 512 bytes
I/O size (minimum/optimal): 512 bytes / 512 bytes
Disk identifier: 0x000c8a7c

   Device Boot      Start         End      Blocks   Id  System
/dev/sda1   *        2048    39845887    19921920   83  Linux
/dev/sda2        39847934    41940991     1046529    5  Extended
/dev/sda3        41940992   125829119    41944064   83  Linux
/dev/sda5        39847936    41940991     1046528   82  Linux swap / Solaris
```

FIGURE 11.5
Example of fdisk.

tracks while also providing the individual number of blocks and file system type for each partition and indicates the bootable partition.

In the Ubuntu Server 14 VM, the software "gparted" [7] can be used to view the existing partitions of the system. The file system of the partitions can be converted to another (for example, "/dev/sda3" from EXT4 to NTFS) by right-clicking on the partition in "gparted" and selecting "Format to" and then choosing a file system, as shown in Figure 11.6.

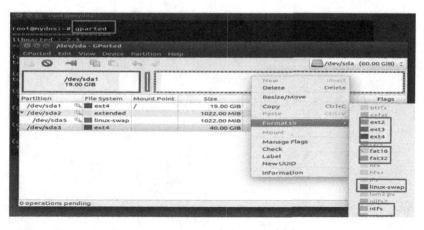

FIGURE 11.6
Example of gparted.

11.5 Hard Disk Analysis in Linux

Various tools can be used to analyse hard disks and find evidence for digital forensics purposes. This section focuses on the wxHexEditor and Active @ Disk Editor tools.

11.5.1 Hard Disk Analysis Using wxHexEditor

This section demonstrates the contents of a partition in Linux using the wxHexEditor tool. The wxHexEditor [4] is a free hex editor for Linux, suited to processing large files and low-level disk data (raw data). To launch the tool, either type "wxHexEditor" in a terminal or click on its icon in Ubuntu Server. The main window that opens is similar to WinHex [5], as shown in Figure 11.7. To create a new file, navigate to "File," "New" and select the file size (as an example, select "1" for the file size, which is in bytes).

The layout of wxHexEditor is similar to WinHex, with three columns, "Offset," "Hex editor," and "Text," as seen in Figure 11.8. You can edit the values in the hex editor and even increase their size (in bytes) by right-clicking on the hex editor column, selecting "insert," and then inserting the number of bytes to be included in the file. Changing the file's size (in bytes) may require you to save the file under a new name. To alter the contents of the file, make sure that the file is in writable mode by navigating to "Options," "file mode," "writable," or "Direct write."

Type "wxHexEditor/dev/sda1" to view the content of the "/dev/sda1" drive in Hex and ASCII formats, as shown in Figure 11.9. This information demonstrates the entire contents of files and directions involved in "/dev/sda1" drive.

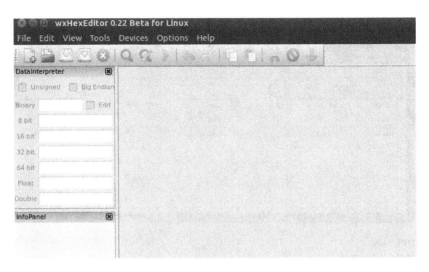

FIGURE 11.7
The wxHexEditor.

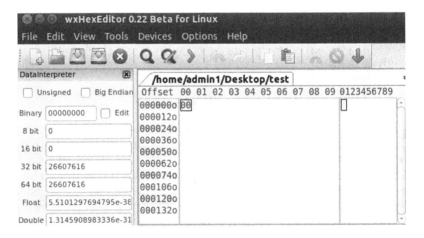

FIGURE 11.8
The wxHexEditor file view.

11.5.2 Crime Investigation: Adding/Changing Files' Content Using wxHexEditor

In the Ubuntu Server, a directory named "Data" is located on the Desktop, including a.pdf file and some images. Perform the following two scenarios.

- **Scenario 1:** Open the "snort-dt.pdf" file located in the "Data" directory in wxHexEditor and copy/paste the hexadecimal value of the word "Digital Forensics" at the beginning of the file ("snort-dt.pdf").

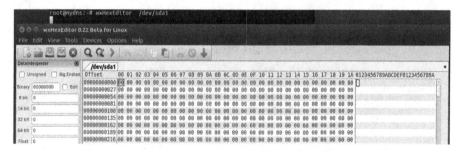

FIGURE 11.9
View of sda1 drive in wxHexEditor.

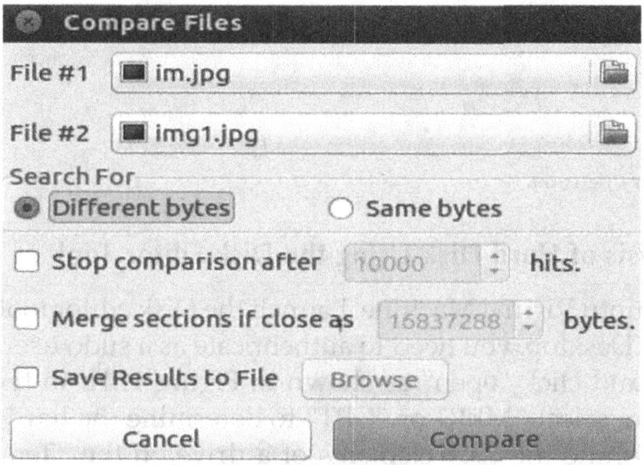

FIGURE 11.10
File comparison with wxHexEditor.

To get the hexadecimal value of the string "Digital Forensics," create a text file with the string as its content and open it in the hex editor. Save the resulting document under a new name ("snort_dt_edited. pdf"). Close the hex editor and directories and access the original and modified files. What do you observe?

- **Scenario 2:** Open the "img1.jpg" file located in the "Data" directory in wxHexEditor and replace the first three bytes in the hex column by "00 00 00" ("snort-dt.pdf"). Save the resulting document under a new name ("img1_edited.jpg"). Close the hex editor and the directories and try to open both the original and the modified file. What do you observe?

To check files for differences in wxHexEditor, similarly to WinHex, navigate to "Tools", "Compare files," select the two files in the "File #1" and "File #2" selectors and select "Different bytes," as shown in Figure 11.10. The changes will be colour-coded in the hex editor columns of the files. You can compare the original "img1.jpg" and the modified "img1_edited.jpg" image files from Scenario 2.

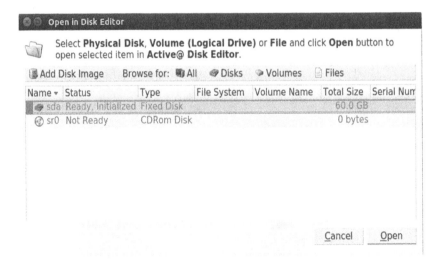

FIGURE 11.11
Using disk editor open sda.

11.5.3 Analysis of Hard Disk Using the Disk Editor Tool

Using the Ubuntu Virtual Machine, Launch the Disk editor tool, found on the left bar of the Desktop, you need to authenticate as a sudo user. Select "Open Disk," "sda" and click "open" as shown in Figure 11.11. In the "Templates" dropdown bar, select "MBR" or "GPT" to determine the hard disk architecture. To determine the EXT elements of a drive, in the "Templates," select "EXT/2/3/4 superblock" and "EXT2/3/4 Inode." The information found in the "Templates" can be found in the column, as shown in Figure 11.12.

Both MBR and GPT partition styles [9,24] have the same properties and elements in Linux OS as in Windows OS. In MBR, the first elements include the reserved code for bootstrap and a disk serial number, followed by the partition table made up of four entries, and then the signature "0x55AA," signifying the end of the MBR. In GPT, the header is followed by 128 partition entries, 128 bytes long. The Ubuntu Server 14 VM that we have been using as an example in this chapter uses MBR, not GPT, as its partition scheme.

11.6 Mount File Systems in Linux

The mounting process involves accessing files on a remote or local drive by attaching the drive to the currently active OS. The default mounting point in Linux OS (including Kali Linux) is a directory called "mnt" found under the root directory. In the Kali VM, you can access the "mnt" directory by

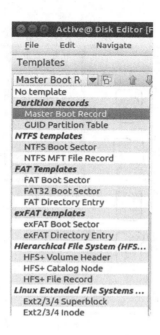

FIGURE 11.12
Disk editor templates.

opening a terminal and typing "cd /mnt," and using the "ls" command to view its contents (mounted devices that appear as directories). As an example, you will mount the "usr" directory of Ubuntu Server VM in Kali.

At this point, the local OS (Ubuntu) can access the mounted partition and a user can use either the absolute or relative path of a file to access it. A full path is the representation of the location of a file that includes all directories, starting from the root directory "/," for example, in Figure 11.13, "/home/cbw/cs5600/hw4.pdf" refers to the pdf under the "cs5600" directory. A relative path is the representation of the location of a file by using the working directory (the directory that we are currently accessing), as the start of the path, for example, if the working directory is "/home/cbs," then the relative path for the hw4.pdf would be "./cs5600/hw4.pdf."

11.6.1 Remote Connection Using SSHFS

Mounting drives over a network is both convenient and possible. For example, in the Kali Linux VM, we can mount a folder from the Ubuntu VM by adhering to the following process. First, open a terminal and move to the "/mnt" directory "cd /mnt," next create a sub-directory, under which the remote folder will be mounted "sudo mkdir new_directory" and finally issue the command "sshfs remote_user@ip:/path/to/remote/dir /mnt/new_directory." The "sshfs" command receives the remote system IP address and a valid username to use to access the remote directory

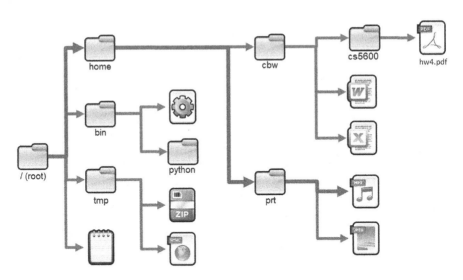

FIGURE 11.13
Example of Linux directory tree.

(specified after the ":" character) and mounts it on a specified directory path ("/mnt/new_directory").

To give an example, go to the "mnt" directory in Kali (by using "cd/mnt") and then create a sub-directory via the command "sudo mkdir server_usr," which will be used as the mounting point. The command used is "sshfs," which is a version of the ssh command used for mounting directories. The command is as follows "sshfs root@192.168.159.152:/usr /mnt/server_usr," which specifies the user that we are using to authenticate as in the remote machine ("root"). This is followed by the IP address of the remote machine, then the directory that we are mounting (after the ":" character, "/usr"), and, finally, the mounting point in the local machine ("/mnt/server_usr"). This method allows you to have swift access to the contents of the remote directory without needing to copy the data to your Kali machine.

To mount a file system, its superblock first needs to be read [17]. A superblock contains information describing a file system, including its meta-data, version, size, key structures on the disk, and more [19]. After the information that characterized the file system has been determined, its mounting point needs to be set. For mounting, in Windows, a letter that characterizes the drive is selected, while in Linux, the new file system is mounted under the directory ("/mnt").

11.6.2 Remote Connection Using SSH

In this section, remote connection to the Ubuntu Server through the Kali Linux VM is explained. To do this, you will be using the SSH command [14].

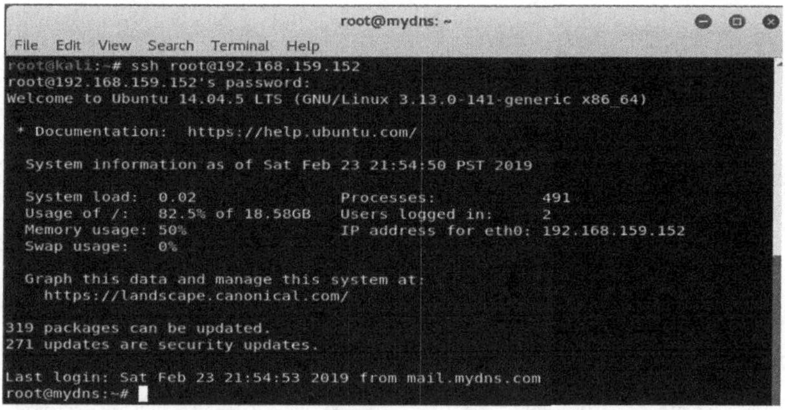

FIGURE 11.14
Example of remote connection using SSH.

The command is structured as follows: "ssh <remote_username>@<remote_IP_address>," where the "<remote_username>" refers to the username of a user that exists in the local machine, while "<remote_IP_address>" is the IP address of the remote machine. For example, to connect to the Ubuntu Server through Kali Linux, the command to connect as the root user would be "ssh root@192.168.159.152," as depicted in Figure 11.14. While connected to the Ubuntu server through ssh, use the "fdisk -l" command to view the available partitions.

11.6.3 Sharing and Mounting Files/Images between Various Virtual Machines

It is vital to share files and directories between various OSs and virtual machines due to the difficulty of copying large data volumes and maintaining data integrity while investigating a cybercrime. For example, copy some files from the Windows 10 VM to the Ubuntu Server VM via FileZilla [3], an application used to transfer files between remote machines of the same or varying OSs, and then you will access those files from Kali by mounting the directory, where these files were copied. First, create a new directory under the path "/home/admin1/windows10_files," which is where files will be copied from Windows.

In Windows, access the FileZilla tool and add the following information (which describes the Ubuntu Server and the user we are login in as), Host: "192.168.159.152," username: "admin1," password: "admin," port: "21" and then click "Quickconnect." If the connection is successful, find the "USB" folder under the "F" partition in Windows and copy its content in the "windows10_files" directory in the Ubuntu Server, as seen in Figure 11.15. You will now be able to locate the transferred ".E01" file in the Ubuntu Server in the "windows10_files" directory.

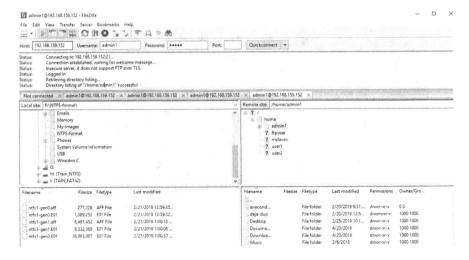

FIGURE 11.15
Example of copying files from Windows to Ubuntu Server using FileZilla.

In Kali, create a new directory under "mnt," where you will mount the Ubuntu Server directory, in which you just copied the ".E01" file. Issue the following commands, first to create the new mounting point (directory), "mkdir /mnt/windows10_files," then mount the remote directory with "sshfs," "sshfs root@192.168.159.152:/home/admin1/windows10_files/mnt/windows10_files." After the mounting process is complete, copy the image (".E01") file to the "output" directory, located in the Desktop ("/home/Desktop/output").

11.7 Data Acquisition in Linux

This section presents the "dd" [13] and "dcfldd" [12] tools to create images in Linux.

11.7.1 The dd Command

The "dd" command is a Unix tool employed in various digital forensic tasks. One such task is the generation of raw images of files, folders, volumes, or physical drives, similar to how FTK Imager [11] is used in Windows. However, unlike FTK Imager, dd does not have a GUI, and it is instead used via a terminal interface in Linux.

In Kali Linux, open a terminal and go to the output directory at Desktop "cd /root/Dekstop/output." To find more information about the "dd" command, type "dd - - help." Next, we will be connecting to Ubuntu Server

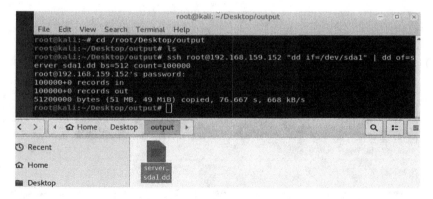

FIGURE 11.16
Example of using the dd command.

and issuing a single "dd" command to read the "/dev/sda1" partition and then piping the output to another "dd" command to save the generated Ubuntu drive forensic image in Kali Linux. The command is as follows "ssh root@192.168.159.152 "dd if=/dev/sda1" | dd of=server_sda1.dd bs=512 count=100000" (enter the root user's password, which is "admin"), as shown in Figure 11.16.

In the previous command, "if" and "of" correspond to "input file" and "output file," respectively, with the first argument used to read the partition and the second ("of") used to output the produced forensic image in the local (output) directory. The "bs" argument refers to the "byte size" that will be read/written at a time, while the "count" refers to the number of "bs" that will be read in total. Note that the file size of the "server_sda1. dd" must be "bs x count" bytes. Thus, only a portion of the "/dev/sda1" drive was copied, where the .dd file is 51.2 MB, and the entire drive is 20.4 GB.

11.7.2 The dcfldd Command

The "dcfldd" command is a Unix tool that functions as an enhanced version of "dd" with more forensic and security features included. Such features include on-the-fly hashing and verification, which is applied to data as it is transferred, ensuring data integrity and status notification. This informs the user about the completion time and the amount of transferred information, multiple, split, and log output, allowing for the hashing of large disks into multiple/ manageable files and saving the hash digests into text files for later comparisons. In Kali Linux, open a terminal and go to the output directory at Desktop "cd /root/Dekstop/output."

To find out more information about the "dcfldd" command, type "dcfldd - - help." Next, we will be connecting to Ubuntu Server and issuing a single "dcfldd" command to read the "/dev/sda1" partition and then

FIGURE 11.17
Example of using the dcfldd command.

piping the output to another "dcfldd" command to save the generated Ubuntu drive forensic image in Kali Linux. As shown in Figure 11.17, the command is as follows "ssh root@192.168.159.152" "dd if=/dev/sda1" | dcfldd of=server_sda_2GB.dd hash=md5 hash=sha1 hashlog=hash_data.txt bs=2048 count=100000" (you will be prompted to enter the root user's password, which is "admin").

The "dcfldd" has similar parameters as "dd" with some exceptions. The "hash" and "hashlog" parameters instruct the tool to create hash digest values of the type specified after "hash" and store them in the "hashlog" file in the local directory. In the above command, the ".dd" file will be equal to 2GB ("bs*count" Bytes). You can verify the integrity of the forensic image that was computed by comparing the hash values that "dcfldd" produces in the log file with values produced by hashing the image with "md5sum" and "sha1sum" tools in the terminal.

11.8 Conclusion

This chapter has discussed the key features of Linux, namely the Kernel, networking, and File Systems. Additionally, the MBR and GPT styles in Linux were also explained. The hard disk analysis in Linux was also described using the wxHexEditor and Active Disk Editor tools. Mounting files and images in Linux were also presented. Finally, the data acquisition in Linux using the dd and dcfldd commands was also explained. Chapter 12 will describe the advanced Linux operations for digital forensics purposes.

References

1. fdisk tool, https://www.computerhope.com/fdiskhlp.htm, 2021.
2. Debian os, https://www.debian.org/, 2021.
3. Filezilla tool, https://filezilla-project.org/, 2021.
4. Fedora os, https://getfedora.org/, 2021.
5. Arch linux, https://archlinux.org/, 2021.
6. Centos, https://www.centos.org/, 2021.
7. gparted tool, https://gparted.org/, 2021.
8. Ubuntu os, https://ubuntu.com/, 2021.
9. Pawan K Bhardwaj, Chapter 6: Managing hard disks with the diskpart utility. *How to Cheat at Windows System Administration Using Command Line Scripts, How to Cheat*, pp. 171–201. Syngress, Burlington, 2006.
10. gnome disk utility, https://packages.debian.org/buster/gnome-disk-utility, 2021.
11. ftk imager tool, https://accessdata.com/product-download/ftk-imager-version-4-2-1, 2021.
12. dcfldd, https://www.linuxnix.com/what-you-should-know-about-linux-dd-command/, 2021.
13. dd, https://www.linuxnix.com/what-you-should-know-about-linux-dd-command/, 2021.
14. ssh, https://www.geeksforgeeks.org/ssh-command-in-linux-with-examples/, 2021.
15. Sayed Naem Bokhari. The linux operating system. *Computer*, 28(8):74–79, 1995.
16. Alireza Shameli-Sendi. Understanding Linux kernel vulnerabilities. *Journal of Computer Virology and Hacking Techniques*, 17(4):1–14, 2021.
17. Brian Carrier. Digital crime scene investigation process. In: *File System Forensic Analysis*, pp. 13–14. Addison-Wesley Professional, Boston, MA, 2005.
18. Terry Collings and Kurt Wall. *Red Hat Linux Networking and System Administration*. Wiley, Hoboken, NJ, 2007.
19. Robert Cowart and Brian Knittel and. *Special Edition Using Microsoft Windows XP Home*. Special Edition Using Series. Que, London, 2004.
20. Brendan Choi. Creating an Ubuntu server virtual machine. *Introduction to Python Network Automation*, pp. 169–222. Apress, Berkeley, CA, 2021.
21. Michael Kerrisk. *The Linux Programming Interface: A Linux and UNIX System Programming Handbook*. No Starch Press, San Francisco, CA, 2010.
22. Wolfgang Mauerer. *Professional Linux Kernel Architecture*. Wiley, Hoboken, NJ, 2010.
23. Andrew Minteer. *Analytics for the Internet of Things (IoT)*. Packt Publishing, Birmingham, 2017.
24. Bruce J Nikkel. Forensic analysis of GPT disks and guid partition tables. *Digital Investigation*, 6(1–2):39–47, 2009.
25. Mario Santana. Chapter 6: Eliminating the security weakness of Linux and Unix operating systems. In: John R Vacca (editor), *Network and System Security* (Second Edition), pp. 155–178. Syngress, Boston, MA, 2014.
26. Sameer Seth and Ajaykumar Venkatesulu. *TCP/IP Architecture, Design, and Implementation in Linux*. Practitioners. Wiley, Hoboken, NJ, 2009.

27. Mark G Sobell and Peter Seebach. *A Practical Guide to UNIX for Mac OS X Users*. Pearson Education, London, 2005.
28. William von Hagen. *Ubuntu 8.10 Linux Bible*. Wiley, Hoboken, NJ, 2009.
29. Matt Welsh, Matthias Kalle Dalheimer, Terry Dawson, and Lar Kaufman. *Running Linux*. Essential Guide to Linux Series. O'Reilly, Sebastopol, CA, 2003.
30. Jason Williams, Peter Clegg, and Emmett Dulaney. *Expanding Choice: Moving to Linux and Open Source with Novell Open Enterprise Server*. Novell Press. Pearson Education, London, 2005.

12

Advanced Linux Forensics

12.1 Introduction

In the previous chapter, we discussed the Linux OS and its kernel, networking, and file systems. This chapter explains EXT and its layout. Then, it further elaborates on the characteristics of EXT versions 2, 3, and 4. Finally, we will delve into the forensic implications of investigating EXT file systems in a Linux environment.

The main objectives of this chapter are as follows:

- Discuss the layout of EXT file systems, including Superblocks, Inodes, and data blocks
- Explain EXT2, 3, and 4 file systems
- Learn forensics implications for Linux systems
- Apply Linux forensics analysis and presentation using the Disk editor and Autopsy tools

12.2 Examining File Structures in Linux

In Linux, everything is either a file or a directory, with even directories represented as a type of file. Commands issues by users in a terminal are called in a case-sensitive mode, which means that, for example, "ls," "Ls," "lS," and "LS" are considered four different strings, with only the first ("ls") being a legitimate command to list the contents of a directory [11,17]. The previous chapter states that Linux directories and files are organized in a tree-like structure, with the root directory ("/") placed at the top of the hierarchy.

Directories under the root directory, called top-level directories, maintain important files in a Linux distribution, with the core top-level directories being "/usr," "/home," "/etc," "/root," "/dev," "/var" [13]. The "/usr" directory and its subdirectories "/bin" and "/sbin" store most applications and commands (most in binary form). The "/home" directors the home directory for every user in the system, with the directory, often named after the user. System configuration files, editable via a text editor, can be found in "/etc," while "/root" is the sudo user's home directory. In the "/dev" directory, several files that function as interfaces for connected devices can be found. For instance, "/dev/sda" could be a storage media disk to which the letter "a" has been assigned (last character of the name). The "/var" directory is used to store several varied, and often important for a forensic investigation, types of files, such as system logs, emails, printing spool directories, and temporary files.

Some of these top-level directories house important files that can assist forensic investigators in identifying useful traces or evidence. In the "/etc" directory, important files are "fstab," "shadow," "group," and "passwd" [10,15]. The "/etc/fstab" file maintains a table of the mounted devices/partitions, file systems, and mounting points. It is read by the "mount" command, either during boot time or when the user issues a modification to the mounted devices. The "/etc/shadow" is considered the master password file, restricted to the root user and the "shadow" group. It maintains the hash values for all local user accounts' passwords and several password-related maintenance fields (expiration dates, renew days). The "/etc/group" file stores information about group membership for all local users. The "/etc/passwd" file contains information about each user account, like username, password, user identifier set by the OS and home directory.

In the "/var" directory, the important files under "log" ("/var/log") are "lastlog," "wtmp," "dmsg," and "syslog," and under "run" ("/var/run"), "utmp" [10,11]. The "/var/log/lastlog" file is a log that stores the last time that each user logged into the system, with each entry including the name of the user, the time of the login that the port that was used. The "/var/log/wtmp" file maintains records of all logins and logouts for the users of a system, in contrast to "/var/run/utmp," which maintains the login information for all currently logged-in users. The "/var/log/dmsg" file is a system message log that maintains information related to the status of hardware connected to the system and is used as a ring buffer (old data is overwritten by new when a file is full). The "/var/log/syslog" file, often called "system.log" or "kernel.log," stores activity data, general and diagnostic messages from the entire system.

12.3 Generic Linux File System Layout (EXT2, 3, 4)

Although Linux OS is compatible with multiple file systems, the default choice for many distributions is EXT, developed in 1992 specifically for Linux OS [14]. Newer versions of the EXT file system include EXT2, 3, and 4, with each newer version being backwards compatible, supporting from 4 Terabytes up to 1 Exabyte and including a journal in versions 3 and 4 [12].

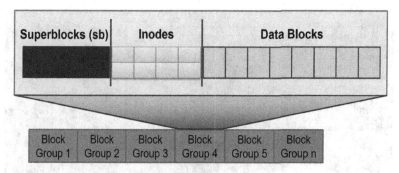

FIGURE 12.1
Structure of block groups of a partition in EXT.

In EXT, each partition is separated into two classes of blocks of data, the boot blocks and the block groups [8]. Each partition has at most one boot block, which stores the necessary information for the proper boot process of the OS stored in a partition. The rest of the partition stores multiple block groups, which themselves are made up of three types of blocks, the Superblocks, the Inode list, and the data (including both files and directories in this category) [11], as seen in Figure 12.1. The Superblocks store a file system's metadata and all the necessary information about how the file system is configured. The information involves parameters as block size, block address range, mount status, the total number of Inodes and blocks, how many are free, and how many are allocated to each block group [8,11].

As the Superblock is integral to access the file system, and thus to protect the integrity of the file system, redundancy has been sought by copying it on various locations of the partition. In the first version of EXT2, a copy of the Superblock was stored at the beginning of every block group. Later versions opted for reducing the produced overhead from this practice by choosing a subgroup of blocks to store backups. The Inodes (Index-nodes) store the file/directory metadata (permissions, owner, size, access/change time), including all their attributes, and maintain a map that links disk blocks with their corresponding file [19]. Each Inode corresponds to a file/directory, and there exists an Inode table in each block group that stores Inodes (files). The data segment of a block group is a collection of blocks that maintains the actual content of files addressed and referenced by the Inode.

12.4 Accessing Block Group Information in Linux

You can view information about the block groups in the Ubuntu Server by using the mke2fs tool [6], as shown in Figure 12.2. In the command displayed in Figure 12.2, the "-b" parameter sets the block size to "1024," followed by

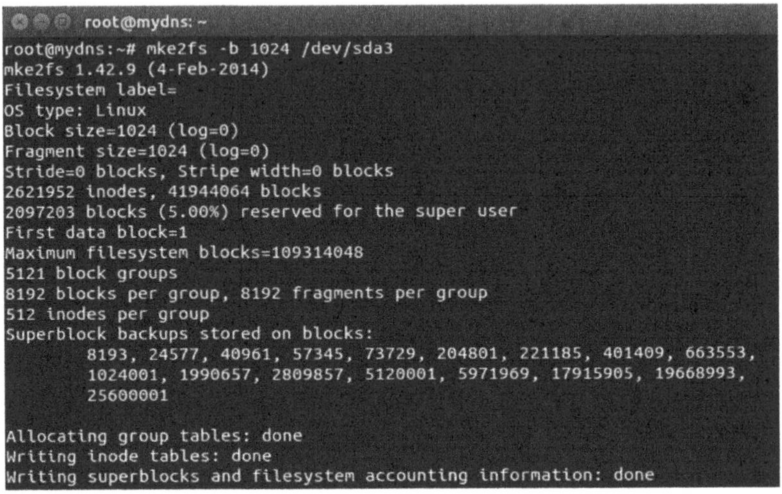

```
😣 😐 😑  root@mydns: ~
root@mydns:~# mke2fs -b 1024 /dev/sda3
mke2fs 1.42.9 (4-Feb-2014)
Filesystem label=
OS type: Linux
Block size=1024 (log=0)
Fragment size=1024 (log=0)
Stride=0 blocks, Stripe width=0 blocks
2621952 inodes, 41944064 blocks
2097203 blocks (5.00%) reserved for the super user
First data block=1
Maximum filesystem blocks=109314048
5121 block groups
8192 blocks per group, 8192 fragments per group
512 inodes per group
Superblock backups stored on blocks:
        8193, 24577, 40961, 57345, 73729, 204801, 221185, 401409, 663553,
        1024001, 1990657, 2809857, 5120001, 5971969, 17915905, 19668993,
        25600001

Allocating group tables: done
Writing inode tables: done
Writing superblocks and filesystem accounting information: done
```

FIGURE 12.2
Output of mke2fs command in Ubuntu.

the identifier of the drive ("/dev/sda3") that is scanned by the command. In general, Inodes are smaller than blocks in size, and groups have fewer Inodes than blocks [11].

To calculate the number of Inode blocks per group (blocks that hold Inode information in a group), determine the number of Inodes in a block, which is calculated by dividing the size of a block by the size of an Inode (Bytes per Block/Bytes per Inode=Inodes per Block). Then you divide the number of Inodes per block group by the Inodes per block and get the Inode blocks per group (Inodes per Group/Inodes per Block=Inode blocks per group).

The number of blocks per group can be calculated by multiplying the block size by 8 (Block size ∗8=Blocks per Group) or viewed in the output of mke2fs, as shown in Figure 12.2. To view Superblock-related and block group information for an EXT drive connected to a Linux environment, the dumpe2fs command [5] can be used. An example for accessing such information for the "/dev/sda3" drive being "dumpe2fs -h /dev/sda3," can be seen in Figure 12.3. Alternatively, the Superblock template in the Disk Editor can be employed.

Each file and directory is represented in an EXT file system as an Inode, where information about the file, the block's location that makes up its content, and permissions are maintained. In the terminal, the "ls -I" command [4] can be utilized to view the index number associated with the Inode of a file, and the "stat" [7] command can be employed to view more metadata about a file. Previously spoke about symbolic and hard links. In association with Inodes, symbolic links form a "shortcut" to a file by creating a new Inode that references the original file name.

Deleting the original file results in the symbolic link becoming useless. Hard links, on the other hand, create a new reference to the original file's

```
○ ○ ⊘   root@mydns: ~
root@mydns:~# dumpe2fs -h /dev/sda3
dumpe2fs 1.42.9 (4-Feb-2014)
Filesystem volume name:    <none>
Last mounted on:           <not available>
Filesystem UUID:           5fd5dfe1-4330-46fe-977a-7ef08d91a6bb
Filesystem magic number:   0xEF53
Filesystem revision #:     1 (dynamic)
Filesystem features:       ext_attr resize_inode dir_index filetype sparse_super
large_file
Filesystem flags:          signed_directory_hash
Default mount options:     user_xattr acl
Filesystem state:          clean
Errors behavior:           Continue
Filesystem OS type:        Linux
Inode count:               2626560
Block count:               10485760
Reserved block count:      524287
Free blocks:               10308654
Free inodes:               2626549
First block:               0
Block size:                4096
Fragment size:             4096
Reserved GDT blocks:       1021
Blocks per group:          32768
Fragments per group:       32768
Inodes per group:          8208
Inode blocks per group:    513
Filesystem created:        Mon May 27 08:20:19 2019
Last mount time:           n/a
Last write time:           Mon May 27 08:20:20 2019
Mount count:               0
Maximum mount count:       -1
Last checked:              Mon May 27 08:20:19 2019
Check interval:            0 (<none>)
Reserved blocks uid:       0 (user root)
Reserved blocks gid:       0 (group root)
First inode:               11
Inode size:                256
Required extra isize:      28
Desired extra isize:       28
Default directory hash:    half_md4
Directory Hash Seed:       de4ba10e-9fc0-4b10-814c-0c8acb58d0c2
```

FIGURE 12.3
Output of dumpe2fs command in Ubuntu.

Inode. Deleting the original file does not affect the hard link, as there is still a reference to the underline Inode, and as long as there is an active reference to an Inode, its contents are maintained. Symbolic links can transcend file system boundaries because they reference a file's name, while hard links reference the underline Inode and thus are limited within file system boundaries. Copying a file creates a new, separate Inode from the original, and therefore changes to the original file are not reflected on the new Inode [11].

Directories are considered to be files, and as such, have their Inodes. Each Inode points to at least one data block, where the contents of files and directories can be found. When a file is accessed, first, the Inode of the directory is sought and accessed to determine its data block. Then, the contents of the directories data block are read, and the file's Inode is identified (this is done for each directory in the path pointing to a file), leading to the file's data block(s) and its content. Each Inode can directly point to 12 data blocks and then retains 3 pointers for indirect referencing of blocks, a graphical representation of which can be shown in Figure 12.4.

The 13th pointer is used to store a pointer to a new block (thus an indirect pointer) that stores pointers to data blocks. The 14th pointer stores a pointer

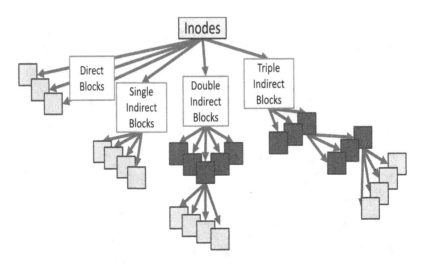

FIGURE 12.4
Inode tree structure and pointer types.

to a block of indirect pointers (therefore a double indirect pointer) that point to data blocks. The 15th pointer stores a pointer to blocks that point to blocks that maintain indirect pointers (thus a triple indirect pointer). The 13th, 14th, and 15th pointers are used when a file increases in size past a pre-defined limit (48KB for 4048 Bytes data blocks).

To specify which Superblock will be used before mounting a partition. For example, to mount "/dev/sda3" in the Ubuntu Server VM, type "mount -o sb= /dev/sda3 /mnt." The "-o" option allows the user to override any parameters about the mounting process. The "sb=" parameter (set to blank) can be used to set a different Superblock than the first one, with copies set every 8192 blocks (first three valid options: 1, 8193, 16385).

Next is the identifier of the partition that will be mounted "/dev/sda3/mnt" followed by the mounting point "/mnt." The newly mounted partition will be visible from Gparted and include "/mnt" as the mounting point. After the drive has been mounted, you can copy the "snort" text file from the VMs Desktop directory to the mounted partition as such: "cp/home/admin1/Desktop/snort/mnt." To unmount the partition, simply use "unmount /dev/sda3."

12.5 EXT File System Versions and Characteristics

12.5.1 EXT2 File System

EXT2 is the second version of the EXT file system developed in January 1993 by Remy Card, designed to provide extended functionality and be extensible

[3]. In EXT2, data is stored in block groups, and bitmaps are used to search and keep track of allocation status for Inodes and blocks. The user can alter the logical block size during the file system creation, with possible values 1024–4096 bytes. The file system provides file names up to 255 characters, with a maximum size of files between 2 GB and 4 TB. Additionally, EXT2 supports fast symbolic links.

When the path to the original file is up to 60 characters, it is stored in the 12 pointers instead of utilizing a new data block. Furthermore, EXT2 and EXT3 partitions can be converted from one version to another, without the need to backup data, while both EXT2 as well as its successor reserves around 5% of the blocks for the root (superuser) user [3,18].

The advantages of the EXT2 file system are that it incorporates improved algorithms that significantly increase its speed and it utilizes additional time stamps to store the last access and modification for Inodes. It also tracks the state of the file system, utilizing a field in the Superblock to indicate if the file system is "clean" or "not clean." EXT2's disadvantages are that it is not a journaling file system, which does not ensure files' integrity and makes it vulnerable to data corruption. Moreover, large files may be stored across more than one block, hindering performance. As the file system becomes more complex, possible corruption of its underline mechanics becomes more probable.

12.5.2 EXT3 File System

EXT3 is the third extended file system, introduced after EXT2 and merged with the mainline Linux Kernel in November 2001 [11]. The EXT3 was developed with several enhancements in mind, compared to the EXT2 file system. The main advantage of EXT3 over EXT2 is that the former is a journaling file system, resulting in improved reliability and ensuring data integrity in an unclean shutdown.

Other features that were introduced in EXT3 are online file system growth that allows the file system to be increased in size while it is mounted and in use without risk of damaging data and the use of the H-Tree indexing for larger directories, which utilizes the hash values of file names as the keys of a B-Tree structure that has directory index blocks as internal nodes (depth 1) and directory entry blocks as leaf nodes (depth 2), allowing for faster lookups [3]. EXT3 supports a maximum file size between 16 GB and 2 TB, while the maximum file system size is 2 and 32 TB, with these values depending on the block size, which can be set between 1 KB and 8 KB as the maximum number of blocks is set to 2^{32}.

The main disadvantage of EXT3 is that it lacks contemporary file system features, such as extents. This reduces the amount of metadata necessary to reference data blocks. It maintains the addresses of the first and last blocks of a continuous sequence of blocks belonging to a file, dynamic allocation of Inodes, and blocks' sub-allocation. This makes use of the slack created when parts of an allocated block are necessary to store the contents of a file, resulting in internal fragmentation.

12.5.3 EXT4 File Systems

EXT4 is the fourth version of the EXT file system and was introduced in October 2008 as an expanded version from EXT3 [10]. Similar to its predecessor, EXT4 is a journaling file system that is backwards compatible with EXT3 and 2, with several new features. First, EXT4 allows for the creation of an unlimited number of subdirectories, unlike EXT3, which imposes a limit of 32000 subdirectories due to the limit of links per Inode [9]. It supports 48-bit block addressing and supports a maximum file system of 1 EB (1 EB=1024 PB, 1PB=1024 TB, 1 TB=1024 GB) and maximum file size of 16 TB (with 4 KB block size).

EXT4 consists of four block types: 1) the first reserved block(s) related to the boot process (boot block), 2) a superblock that records various characteristics of the file system, 3) the Inode list, and 4) the data block. The EXT file systems utilize indirection to support larger file sizes. However, when the file is fragmented, this results in many disk reads. To increase performance, EXT3 and 4 support extents, which are contingent blocks that belong to a single file, which file systems seek to utilize by storing the pointer to the initial block and the length of blocks that belong to that file. With time, fragmentation occurs, and Inodes store multiple extents for a single file (NTFS also uses extents) [16].

Unlike EXT3, EXT4 Inodes include four extent pointers instead of block pointers, with each extent addressing at most 128 MB of contiguous space (for 4 KB blocks). If more space is needed to store a file, a new data block is allocated, similar to how indirect pointer blocks were allocated in previous file system versions. Some advantages of EXT4 over previous versions are that it supports all the basic file system functionality that is required for smooth execution, improves performance from EXT3 by using extents, and optimizes the block group structure combined with a B-tree directory indexing [10]. Some disadvantages of EXT4 are that it is an incremental improvement over EXT3 and lacks some contemporary file system features such as copy-on-write. Table 12.1 lists a comparison between EXT2, 3, and 4.

12.6 Forensic Implications of EXT File Systems

Through commands found in Linux OS, it is possible to recover corrupt blocks and view Inode-related information. To recover corrupted block groups in the Ubuntu Server VM, first, unmount the drive "unmount/dev/sda3" then

TABLE 12.1

Comparison of EXT File Systems

Point	EXT2	EXT3	EXT4
Maximum file size	16 GB to 2 TB	16 GB to 2 TB	16 GB to 16 TB
Maximum file system size	2 TB to 32 TB	2 TB to 32 TB	1 EB
Journaling	Not available	Available	Available (can be disabled)
Number of directories	31998	31998	Unlimited
Journal Checksum	No	No	Yes
Multi-block allocation and delayed allocation	No	No	Yes

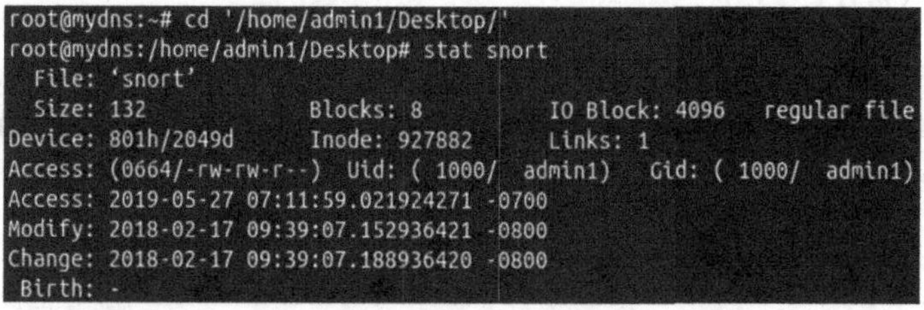

```
root@mydns:~# cd '/home/admin1/Desktop/'
root@mydns:/home/admin1/Desktop# stat snort
  File: 'snort'
  Size: 132            Blocks: 8       IO Block: 4096   regular file
Device: 801h/2049d     Inode: 927882   Links: 1
Access: (0664/-rw-rw-r--)  Uid: ( 1000/  admin1)   Gid: ( 1000/  admin1)
Access: 2019-05-27 07:11:59.021924271 -0700
Modify: 2018-02-17 09:39:07.152936421 -0800
Change: 2018-02-17 09:39:07.188936420 -0800
 Birth: -
```

FIGURE 12.5
Stat command example.

clean and recover the block groups of the drive using the "fsck" command as in the following "fsck-v/dev/sda3." To determine the Inodes of a file, use the "stat" command as shown in Figure 12.5.

To view the index number of files in your current directory, use the "ls-li" command, where the "l" option displays the contents of the directory in list mode and "i" displays the index number. The metadata of a file system can be viewed with the "debugfs" command ("debugfs/dev/sda3").

12.6.1 Case Study: Linux's Accounts

Use a dictionary or brute force attack to crack passwords of all the users of the VM of Ubuntu server 14.

1. What are dictionary attacks and brute force attacks?
2. Discuss the digital evidence steps.
    2.1. How could you crack the user accounts and their passwords? (focus on passwd and shadow files, and you need to unshadow the files).
    2.2. What are the tools and commands used? (you can use the john the Ripper tool on the VM of Kali).

12.7 Data Analysis and Presentation in Linux

This section explains how to utilize the Disk Editor and Autopsy tools for analysing and presenting a forensic image obtained from a Linux VM.

12.7.1 Examining Superblock and Inode Information in Disk Editor

Launch the Disk Editor [1] in the Ubuntu Server VM. Next, select "sda-fixed disk" and then "Local Disk (/dev/sda1)" as shown in Figure 12.6. To determine the Superblock and Inode information of the "etc" directory, first select the directory in the right panel, next click the "Inspect File Record" as presented in Figure 12.7, and from the templates, select "EXT2/3/4 Superblock" or "EXT2/3/4 Inode."

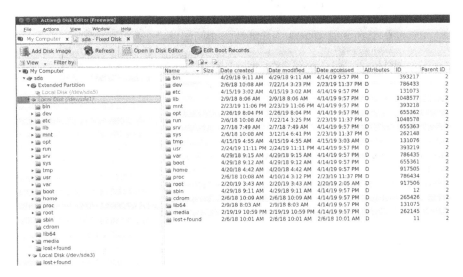

FIGURE 12.6
Disk Editor view Local Disk.

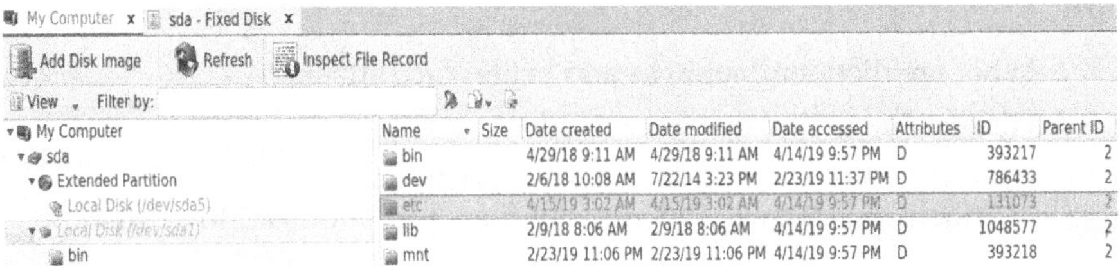

FIGURE 12.7
Disk Editor view Superblock and Inode of the etc directory.

12.7.2 Data Preparation Using Autopsy

This section describes the Autopsy tool [2], a digital forensics platform and a graphical interface for many digital forensic tools, such as the sleuth kit. To use Autopsy, first launch the Kali Linux VM and then navigate to "Applications," "11-Forensics," and then "autopsy." A new terminal launches by clicking the "autopsy" icon, displaying the program's information and the details for opening the Autopsy Forensic Browser, as shown in Figure 12.8. To open the Autopsy browser, right-click on the link in the terminal and select "Open Link."

12.7.2.1 Create a New Case in Autopsy Browser

When the Autopsy browser launches, it displays the screen, as shown in Figure 12.9. To start a new case, click on the "New Case" button. Next, you will be prompted to enter the case details, as presented in Figure 12.10. For

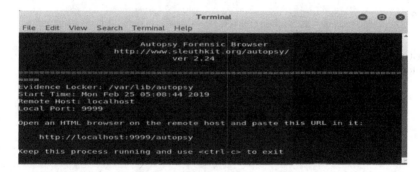

FIGURE 12.8
Autopsy terminal information.

FIGURE 12.9
Autopsy browser initial window.

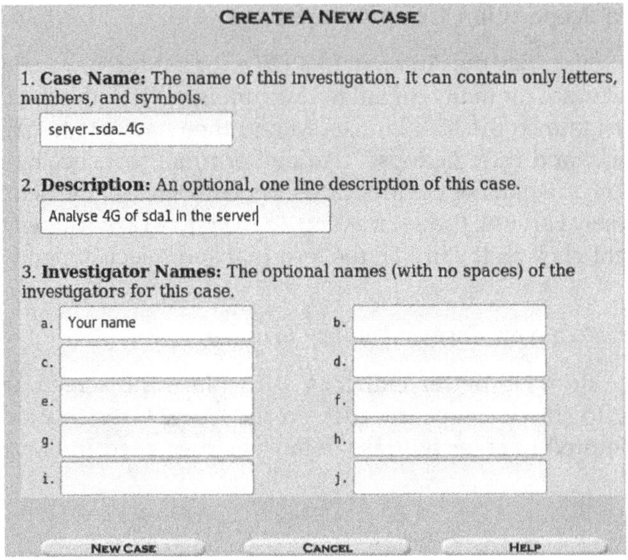

CREATE A NEW CASE

1. **Case Name:** The name of this investigation. It can contain only letters, numbers, and symbols.

    server_sda_4G

2. **Description:** An optional, one line description of this case.

    Analyse 4G of sda1 in the server

3. **Investigator Names:** The optional names (with no spaces) of the investigators for this case.

a.	Your name	b.	
c.		d.	
e.		f.	
g.		h.	
i.		j.	

NEW CASE	CANCEL	HELP

FIGURE 12.10
Autopsy browser new case.

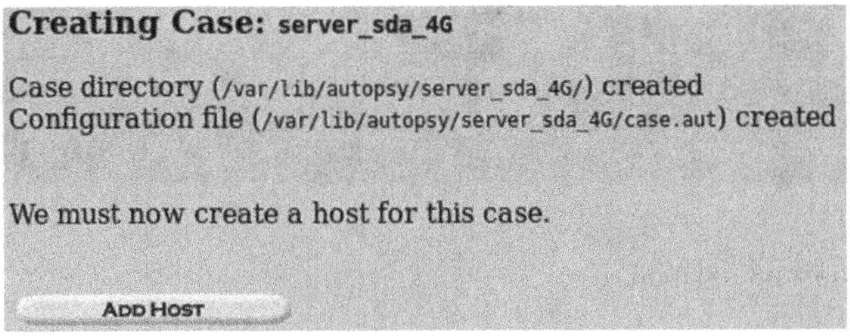

Creating Case: server_sda_4G

Case directory (/var/lib/autopsy/server_sda_4G/) created
Configuration file (/var/lib/autopsy/server_sda_4G/case.aut) created

We must now create a host for this case.

ADD HOST

FIGURE 12.11
Autopsy browser case created window.

this example, enter Case Name: "server_sda_4GB," as you will be analysing the forensic image that was created in the lab of the previous chapter. Several investigator names can be entered, as it is often the case that multiple investigators work on a single case. After entering the requested information, select "New Case." This will prompt Autopsy to create a case directory and a configuration file (with the paths given by the program), as shown in Figure 12.11.

In the "Creating Case" window, select "ADD HOST" and add information for hostname ("ubuntuserver") as shown in Figure 12.12 and select "ADD HOST." After the host has been added and the correct directories have been created, select the "ADD IMAGE" button (Figure 12.13) and add the forensic image that will be analysed. In the new window that appears, select "ADD

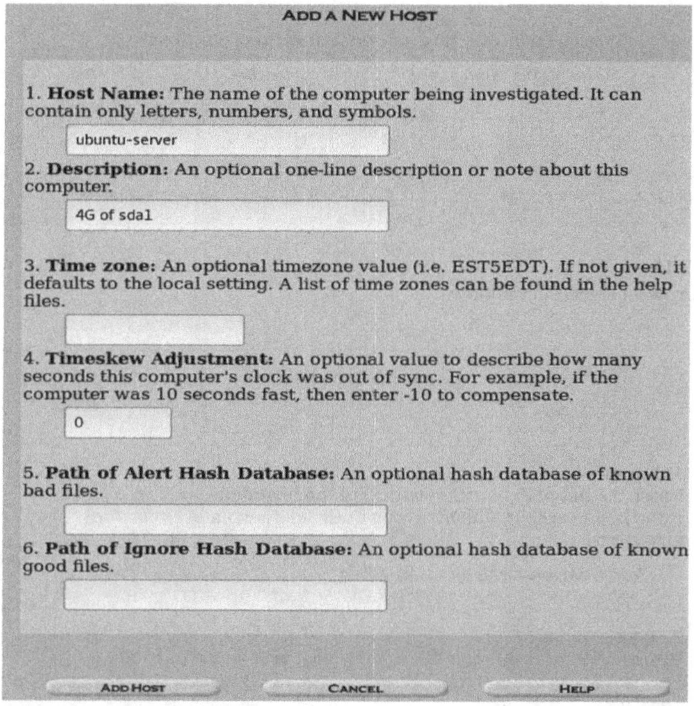

FIGURE 12.12
Autopsy browser adding host window.

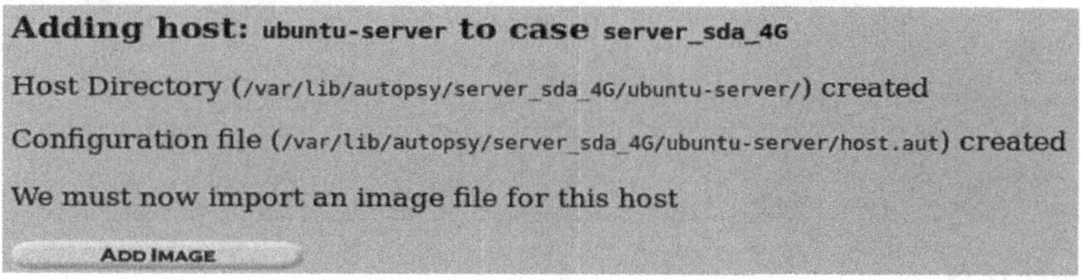

FIGURE 12.13
Autopsy browser host added window.

IMAGE FILE" as in shown Figure 12.14. In the next window, the full path to the image file needs to be inserted, as shown in Figure 12.15, selecting "Partition" as "Type" and "Symlink" as "Import Method," to ensure that the image is imported to the Evidence Locker, without the risks associated with moving or copying the image file. Next, the Image Details for the forensic image that was inserted will be displayed. To verify the integrity of the image, select the "Calculate the hash value of this image" radio button and then select the "Verify hash after importing?."

No images have been added to this host yet

Select the Add Image File button below to add one

ADD IMAGE FILE	CLOSE HOST
HELP	

FILE ACTIVITY TIME LINES	IMAGE INTEGRITY	HASH DATABASES
VIEW NOTES	EVENT SEQUENCER	

FIGURE 12.14
Autopsy browser selection window.

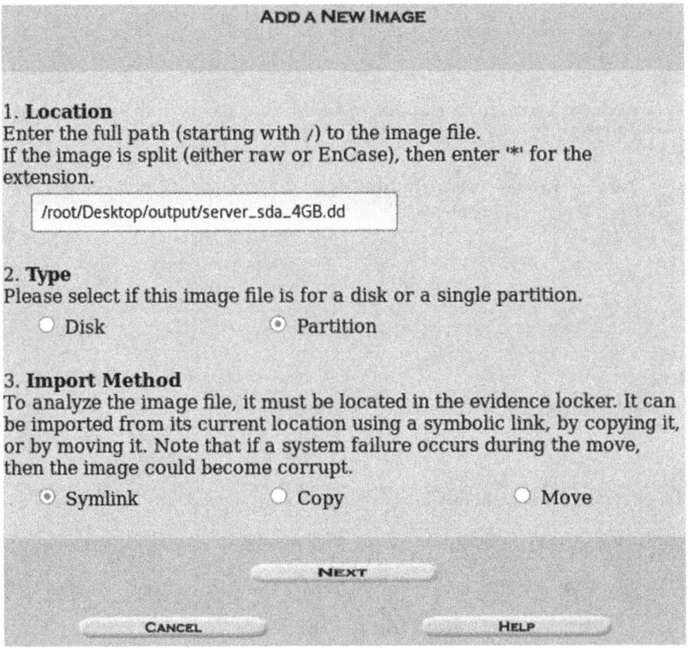

ADD A NEW IMAGE

1. Location
Enter the full path (starting with /) to the image file.
If the image is split (either raw or EnCase), then enter '*' for the extension.

/root/Desktop/output/server_sda_4GB.dd

2. Type
Please select if this image file is for a disk or a single partition.
　○ Disk　　　　　　　⦿ Partition

3. Import Method
To analyze the image file, it must be located in the evidence locker. It can be imported from its current location using a symbolic link, by copying it, or by moving it. Note that if a system failure occurs during the move, then the image could become corrupt.
　⦿ Symlink　　　　　○ Copy　　　　　○ Move

NEXT

CANCEL　　　　　　　HELP

FIGURE 12.15
Autopsy browser import image window.

Additional information about the images will be displayed in this window, as shown in Figure 12.16. After clicking "ADD," Autopsy will import the image and calculate its hash values. You can compare the new values to the ones that were generated when the image was created (saved in the "hash_4GB.txt"). Additional information about the forensic image can be acquired by clicking the "IMAGE DETAILS" option. Before analysing the image, select "Image Integrity" to calculate the MD5 hash of the image. In the new window shown in Figure 12.17, by clicking on the "VALIDATE" button, a new hash value is generated and compared to the original. A match

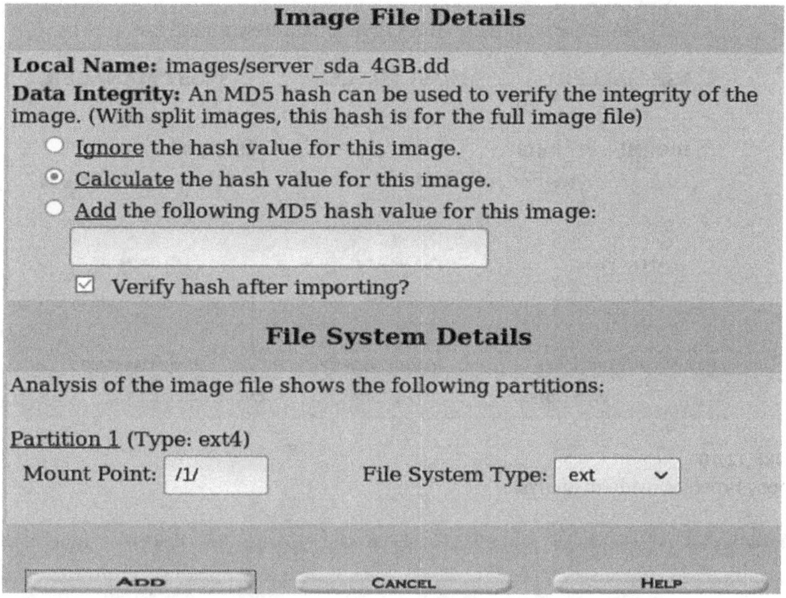

Image File Details

Local Name: images/server_sda_4GB.dd
Data Integrity: An MD5 hash can be used to verify the integrity of the image. (With split images, this hash is for the full image file)

- ○ <u>Ignore</u> the hash value for this image.
- ◉ <u>Calculate</u> the hash value for this image.
- ○ <u>Add</u> the following MD5 hash value for this image:

☑ Verify hash after importing?

File System Details

Analysis of the image file shows the following partitions:

<u>Partition 1</u> (Type: ext4)

Mount Point: `/1/`   File System Type: `ext ∨`

| ADD | CANCEL | HELP |

FIGURE 12.16
Autopsy browser image information window.

FILE SYSTEM IMAGES

server_sda_4GB.dd   F898A3B3C2EFC168E7464A8B27BF29A5 | VALIDATE |

| CLOSE | REFRESH | HELP |

FIGURE 12.17
Autopsy browser validation window.

indicates that the file has not been modified in the process. To start the analysis, click on "ANALYSE," as shown in Figure 12.18.

12.8 Case Analysis Using Autopsy

Analysis can occur after a new case has been created and a forensic image imported to the case Evidence Locker. The first window in the analysis process presents the investigator with multiple choices, as shown in Figure 12.19. General details about the image that will be investigated can be accessed by selecting "IMAGE DETAILS" (Figure 12.20). Clicking on the "FILE

Select a volume to analyze or add a new image file.
CASE GALLERY
HOST GALLERY
HOST MANAGER
mount
name
fs type
/1/
server_sda_4GB.dd-0-0
ext
details
ANALYZE
ADD IMAGE FILE
CLOSE HOST
HELP
FILE ACTIVITY TIME LINES
IMAGE INTEGRITY
HASH DATABASES
VIEW NOTES
EVENT SEQUENCER

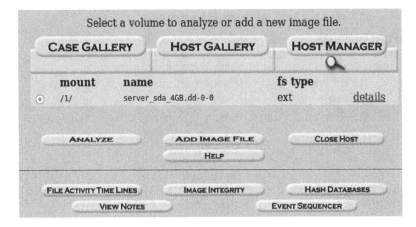

FIGURE 12.18
Autopsy browser ready to analyse.

FILE ANALYSIS
KEYWORD SEARCH
FILE TYPE
IMAGE DETAILS
META DATA
DATA UNIT
HELP
CLOSE
?
X
To start analyzing this volume, choose an analysis mode from the tabs above.

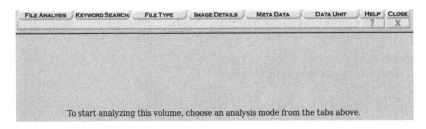

FIGURE 12.19
Autopsy browser analysis mode.

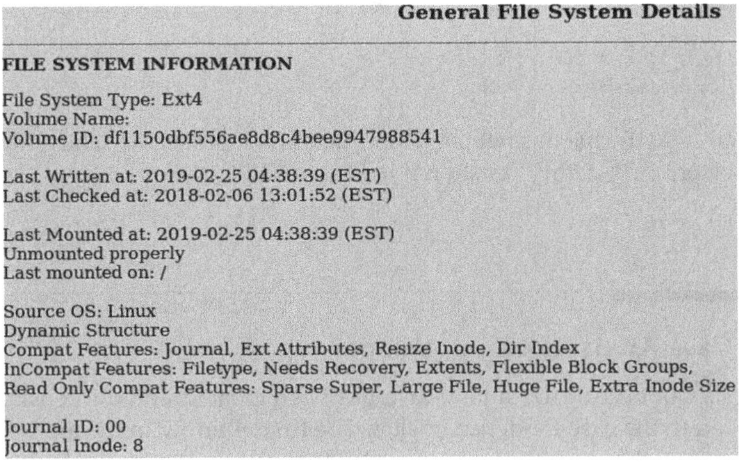

General File System Details
FILE SYSTEM INFORMATION
File System Type: Ext4
Volume Name:
Volume ID: df1150dbf556ae8d8c4bee9947988541
Last Written at: 2019-02-25 04:38:39 (EST)
Last Checked at: 2018-02-06 13:01:52 (EST)
Last Mounted at: 2019-02-25 04:38:39 (EST)
Unmounted properly
Last mounted on: /
Source OS: Linux
Dynamic Structure
Compat Features: Journal, Ext Attributes, Resize Inode, Dir Index
InCompat Features: Filetype, Needs Recovery, Extents, Flexible Block Groups,
Read Only Compat Features: Sparse Super, Large File, Huge File, Extra Inode Size
Journal ID: 00
Journal Inode: 8

FIGURE 12.20
Autopsy browser image details.

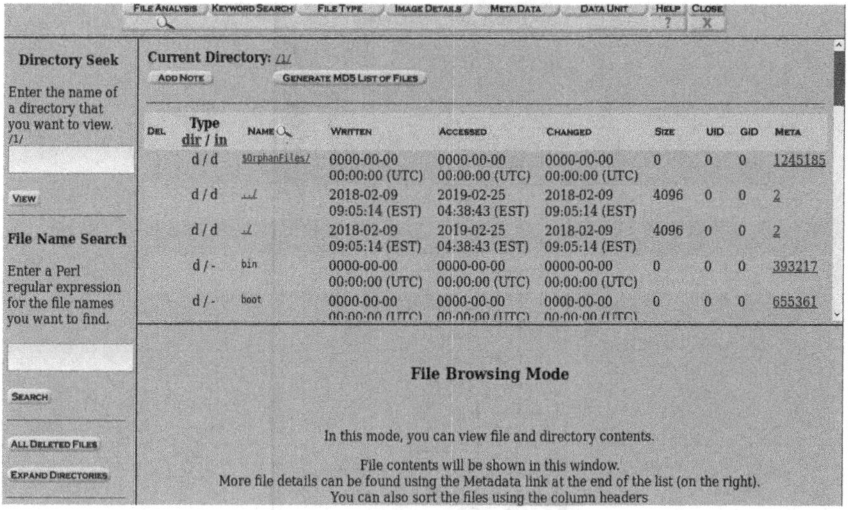

FIGURE 12.21
Autopsy browser file analysis.

ANALYSIS" tab engages the File Browsing Mode (Figure 12.21), allowing the examination of directories and files that exist within the imported image.

Each file and directory has fields that provide information about when they were "WRITTEN," "ACCESSED," "CHANGED," and "CREATED," along with their size and metadata. To ensure integrity can be verified at any stage of the investigation, the button "Generate MD5 List of Files" can be used to produce hash values for all files present in the image that is investigated.

Additionally, investigators can add notes by using the "ADD NOTE" button. By clicking the "EXPAND DIRECTORIES" button, the contents of directories can be more easily viewed, with the "+" character indicating that a directory can be further expanded, as depicted in Figure 12.22. To view deleted files, click on "ALL DELETED FILES." Deleted files are marked in red, and the same fields are used as with regular files ("WRITTEN," "ACCESSED," "CHANGED," etc.).

12.8.1 Sorting Files

A forensic image may include multiple files, which renders the inspection of their metadata inefficient. For that reason, the "FILE TYPE" feature of Autopsy can be used. Through this option, allocated, unallocated and hidden files can be investigated and sorted by type through the "Sort Files by Type." After the sorting process is completed, a "Results Summary" is displayed in Figure 12.23, with some mismatches detected. These mismatches have to be investigated by viewing their metadata, with notes added by the investigators.

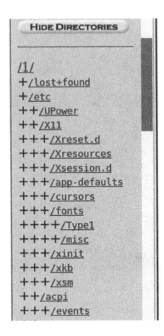

FIGURE 12.22
Autopsy browser add note.

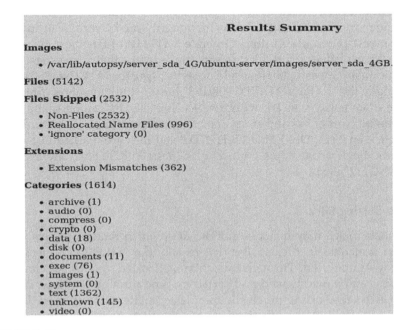

FIGURE 12.23
Autopsy browser results summary.

12.9 Conclusion

This chapter has explained the layout of EXT file systems in Linux. Then, the EXT2, 3, and 4 file systems are discussed. The forensics implications in Linux were also discussed. Data analysis and presentation using the Disk Editor and Autopsy tools were also demonstrated. Next, network forensics will be discussed.

References

1. Disk editor tool, https://www.disk-editor.org/index.html, 2021.
2. Autopsy tool, https://www.sleuthkit.org/autopsy/desc.php, 2021.
3. EC-Council. *Computer Forensics: Hard Disk and Operating Systems*. Cengage Learning, Boston, MA, 2009.
4. Ls linux command, https://linuxize.com/post/how-to-list-files-in-linux-using-the-ls-command/, 2021.
5. dumpe2fs linux command, https://www.geeksforgeeks.org/dumpe2fs-command-in-linux-with-examples/, 2021.
6. mke2fs linux command, https://linux.die.net/man/8/mke2fs, 2021.
7. stat linux command, https://linuxize.com/post/stat-command-in-linux/, 2021.
8. André Årnes, ed. *Digital forensics*. John Wiley & Sons, Hoboken, NJ, 2017.
9. Alfred Basta, Dustin A Finamore, Nadine Basta, and Serge Palladino. *Linux Operations and Administration*. Networking (Course Technology, Inc.). Cengage Learning, Boston, MA, 2012.
10. Richard Blum and Christine Bresnahan. *Linux Command Line and Shell Scripting Bible*. Wiley, Hoboken, NJ, 2015.
11. Brian Carrier. Digital crime scene investigation process. In: *File system forensic analysis* pp. 13–14. Addison-Wesley Professional, Boston, MA, 2005.
12. Vlado Damjanovski. Chapter 9: Video management systems. In: Vlado Damjanovski, (editor) *CCTV* (Third Edition), pp. 322–359. Butterworth-Heinemann, Boston, MA, 2014.
13. Richard Fox. *Linux with Operating System Concepts*. CRC Press, Boca Raton, FL, 2014.
14. Andrew Hoog. Chapter 4: Android file systems and data structures. In: Andrew Hoog (editor), *Android Forensics*, pp. 105–157. Syngress, Boston, MA, 2011.
15. Runkai Yang, et al. Performance modeling of Linux network system with open vSwitch. *Peer-to-Peer Networking and Applications* 13(1):151–162, 2020.
16. Sabo Sambath and Egui Zhu. *Frontiers in Computer Education*. Advances in Intelligent and Soft Computing. Springer, Berlin Heidelberg, 2012.
17. Roderick W Smith. *Linux Samba Server Administration: Craig Hunt Linux Library*. Craig Hunt Linux Library. Wiley, Hoboken, NJ, 2006.
18. Sander van Vugt. *The Definitive Guide to SUSE Linux Enterprise Server*. Apress, Berkeley, CA, 2007.
19. Brendan Choi. Linux fundamentals. In: *Introduction to Python Network Automation*. Apress, Berkeley, CA, 2021. 275–327.

13

Network Forensics

13.1 Introduction

This chapter introduces the domain of network forensics (NF). Initially, we discuss the Open System Interconnection (OSI) model and the Transmission Control Protocol/Internet Protocol (TCP/IP) protocol suite to gain a base understanding of how networks work. Then, the primary stages of networking and NF will be presented, describing the process of discovering, collecting, examining, and analysing network-derived data to reconstruct events and draw conclusions of cyberattacks. Finally, some aspects of practically applying NF will be demonstrated to gain insights into the actual NF process.

The main objectives of this chapter are as follows:
- Discuss the basics of NF
- Learn fundamentals of OSI model and TCP/IP stack
- Explain the stages of NF
- Get insights into the practice of NF

13.2 What Is Network Forensics?

Recall that digital forensics (DF) is a discipline that focuses on investigating cybercrimes, prioritizing the identification of traces and extraction of evidence from a multitude of sources, including hard drives, CPUs, and RAM [6]. With the proliferation of the internet and networking to interconnect remote systems and the expansion of the Internet of Things (IoT), cyber-criminals advanced their attacks to target specific parts of systems and networks, driving the need for specialized DF sub-disciplines, such as NF. NF is a sub-discipline of DF tasked with acquiring, preserving, examining, analysing, and presenting network-related data. This can take the form of either file such as logs stored in a server, or traffic in the form of network

DOI: 10.1201/9781003278962-13

packets, for intelligence gathering, intrusion detection, or as a source of evidence in criminal investigations [7,10].

NF deals with short-lived (volatile) data in motion, such as packets generated by hosts and swiftly transmitted over the network. As such, it is often employed as a proactive process in investigations. It can be utilized in two primary scenarios [5,10]:

- **Security monitoring**: For security purposes, a network can be monitored for anomalous behaviour, leading to identifying intrusion events. An attacker might be able to erase all log files on a compromised host; network-based evidence might therefore be the only evidence available for forensic analysis.

- **Criminal investigation**: For law enforcement, already captured network traffic may be analysed to identify and reassemble transferred files, searching for keywords and parsing human communication such as emails or chat sessions. Two systems are commonly used to collect network data; a brute force "catch it as you can" and a more intelligent "stop look listen" method.

In the greater scope of DF investigations, NF can be viewed as an additional source of evidence that can be employed to collect data and supplement other DF sub-disciplines such as hard drive, CPU, RAM, cloud, and mobile forensics [9]. NF experts are required to have a working understanding of the entire TCP/IP protocol stack, including the communication protocols that span the entire protocol stack (from physical to application layer) that they might need to analyse, the way these protocols process data that is transmitted between hosts, and the inter-layer interactions.

Although experts in NF need to understand multiple protocols and have a working knowledge of several tools that automate some aspects of the investigation (such as collection, preservation, and examination), NF has a low barrier of entry as it is a well-documented discipline with multiple online resources and readily available tools at low-cost, while practitioners can set up virtual machines (VMs) as isolated laboratories for practice.

13.2.1 Benefits and Challenges of Network Forensics

In a world dominated by internet-enabled services and interconnected devices, hackers have harnessed insecure connections and vulnerable devices to launch massive cyberattacks that can affect millions of computers and even cripple the internet [8]. Under these circumstances, the benefits of mastering and employing NF methods become evident. To begin with, it is an additional source of data for the forensic and security experts that, if analysed properly, can complement other data sources and help combine traces from multiple sources and build a more "spherical view" of security events. Furthermore, by investigating networks and network-derived data such as log packets and network flows that describe network interactions between hosts, traces and evidence can be extracted that remotely identify individuals of interest that may have gone unnoticed otherwise. Finally, it is a good document and supported field of DF.

Although the benefits of mastering and employing NF in an investigation or as a continuous monitoring process are substantial, it is vital to consider the challenges of using NF methods and techniques [7]. To begin with, a considerable proportion of NF investigation methods rely on the acquisition and analysis of network traffic to identify traces of cyberattacks, which is where the first challenges arise [9,11]. Network traffic is considered a volatile and short-lived source of data, as traffic is swiftly generated by one host and sent to its intended recipient, often segmented and organized into smaller units of information. It disappears from the transmission medium. Thus, proactive network captures are required for effectiveness, which is expensive.

Additionally, due to the proliferation of cyberattacks in recent years, internet communications often employ encryption schemes to ensure their confidentiality and integrity. However, NF experts cannot easily analyse encrypted traffic, who often need to analyse the payload of exchanged packets to identify evidence of malicious attacks and extract files of interest [11]. An example of encrypted traffic using the IPsec Encapsulated Security Payload can be seen in Figure 13.1. Furthermore, with the number of devices directly connected to the internet continuously rising [3] (one driving factor of which is the IoT), the number of data sources that NF investigators

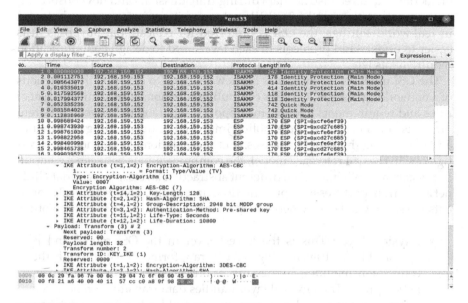

FIGURE 13.1
Wireshark example of encrypted traffic using IPSec.

are required to identify and source from are increasing in size, resulting in added analysis noise which, coupled with a common lack of ground truth, leads to hindrances that negatively affect the effectiveness of investigations.

Added to this, covert side-channels that allow attackers to bypass existing intrusion detection techniques and stealthily communicate with compromised devices, combined with steganography techniques, where attackers hide data in otherwise benign files by making near-imperceivable changes to the original files (such as altering pixel values in images) further inhibit the detection of evidence in network data.

13.3 Networking Basics

In this section, basic networking concepts are discussed. First, the OSI model is explained, followed by the TCP/IP stack of protocols.

13.3.1 Open System Interconnection (OSI) Model

When discussing and analysing network transactions to detect cybersecurity incidents, it is crucial to employ common terminologies and definitions to communicate findings effectively. To that effect, the OSI model is often used as an abstract model for teaching how network interactions break down and reconstruct data during transmissions and how hosts communicate [1]. Although the TCP/IP protocol suite is more commonly found in real-world settings, the OSI is preferred for educational purposes. It is easier to understand protocols in each layer used to design communication requirements [2].

The OSI model is a conceptual framework used to describe the functions of a networking system. It unifies computing functions into a universal set of rules and requirements in order to support inseparability between different products and software while enabling improved understanding of activities and supporting the search, analysis, and presentation of evidence for investigators. In the OSI reference model, the communications between a computing system are split into seven different abstraction layers: physical, data link, network, transport, session, presentation, and application. Figure 13.2 presents the seven OSI layers when two machines communicate with each other.

- **Physical layer:** This is the lowest layer in the OSI model and is responsible for transmitting data in raw format (bits) through the connection medium, harnessing electrical, optical, or electromagnetic signals. The physical layer handles data transmission between neighbouring network elements, such as end-hosts, routers, switches, and bridges.

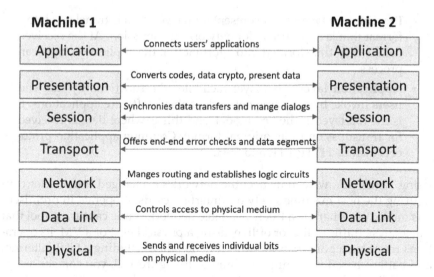

FIGURE 13.2
The OSI model between two machines.

- **Data link layer:** Data is received from higher layers and organized into frames transmitted between neighbouring nodes. Any errors during transmission in the physical layer are also handled in the data link layer. For identifying the next destination in subsequent hops, the Medium Access Control addresses.

- **Network layer:** The network layer groups the datagrams and segments (data from the Transport Layer) into packets and transmits them between sender and receiver. Routing occurs on this layer, where the source and destination identifiers are employed to determine routes through which the packets will be transmitted. The most prominent network layer protocol is the IP, and devices commonly operating on this layer include routers and gateways.

- **Transport layer:** The transport layer produces datagrams and segments, depending on the protocol employed in this layer, and is responsible for end-to-end delivery of data from higher levels. Furthermore, the layer defines the appropriate size and sequencing of data for reconstruction or re-transmission. The most prominent protocols of this layer include TCP and the User Datagram Protocol.

- **Session layer:** The session layer handles the complete end-to-end conversation between two hosts. This layer is responsible for setting up, managing, and terminating a session between two hosts. In connection disruptions, the session layer handles the re-connection and continuation of data transmission without restarting the transmission from the beginning.

- **Presentation layer:** In the presentation layer, data is translated into a format that is appropriate for network transmission. At the receiver, data is then transformed into a format that the application layer can process.
- **Application layer:** This layer facilitates the functionality that end users invoke by accessing network-enabled software applications. It is the top layer in the OSI model, and this is where data is received for transmission in its original format. Common application protocols include HTTP, FTP, and SSH.

Using Wireshark, we can inspect the network's exchanged traffic after configuring the host machine's network interface cards to operate in promiscuous mode. Wireshark is a freely available network traffic capturing tool that can analyse traffic online or offline, using a physical host or a VM. Its operations resemble tcpdump, with extra functionality including a GUI, filtering, and pattern-based searching capabilities. In Figure 13.3, you can see a captured ARP packet analysed into its OSI component layers, with those being the physical, data link, and network layers.

13.3.2 TCP/IP Protocol Stack

Familiarisation with the TCP/IP protocol stack is essential to a forensic expert, prevalent in today's internet. Unlike the OSI, the TCP/IP defines five layers, a physical and data link layer that is equivalent to the OSI's bottom

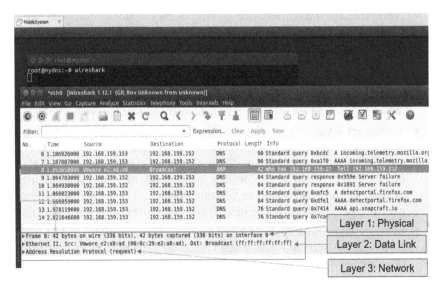

FIGURE 13.3
Wireshark example of ARP packet.

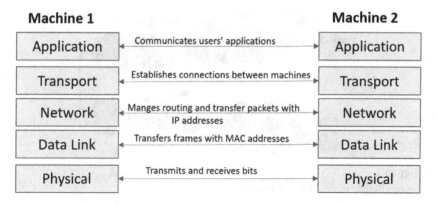

FIGURE 13.4
TCP/IP stack and interactions.

two layers, the internet and transport layers that correspond to network and transport layers in the OSI, respectively, and the application layer that cumulates the top three layers of OSI [2]. Figure 13.4 provides a visual representation of the TCP/IP layers and their interconnection between two machines. Essentially, their functionality is similar to their OSI counterparts, with top layers requesting services from lower layers and data gradually encapsulated as it traverses the layers towards the physical medium.

The importance of the TCP/IP and OSI models becomes evident by considering the tasks of an NF expert. Although the OSI model is generic, it is often employed as a teaching tool to understand basic networking concepts before moving on to the TCP/IP model, a stack of specific protocols. An expert may face multiple protocols employed by malicious actors during an investigation to launch their cyberattacks. These attacks may employ a particular combination of packet flags and fields that, when reviewed in the sequence of exchanged traffic in a suspect network, could indicate an unusual interaction between network entities. Thus, understanding how the protocols in the TCP/IP work under normal circumstances could lead to the detection and identification of cyberattacks, their target, compromised machines, employed methods, the identity of the attackers.

13.4 Network Forensic Investigations

NF is a specialized sub-discipline of DF employed to investigate security incidents that partially or fully occur in a network. Over the years, several investigation frameworks have been proposed for DF, designed for specialized settings (networks, cloud, IoT, mobile) and altering the activities that experts need to carry out in each stage. By comparing this framework,

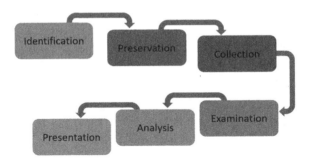

FIGURE 13.5
The stages of the NF investigation process.

standard functionality leads to six stages that define actions shared between most frameworks [9]. Figure 13.5 illustrates the stages of the NF investigation process.

The common stages of the network forensic investigation process include:

- **Identification** is the first stage during an investigation. At first, a security incident needs to be identified for the forensic investigation process to be invoked. Potential sources of relevant information (devices) need to be identified and recorded.

- **Preservation** is the second stage of the NF investigation process. After the devices that will be accessed for data extraction have been identified, it is imperative to ensure the integrity of the data we collect. In other words, investigators need to limit the alterations they impose on the data they extract. For this purpose, specially configured machines that do not shut down computers do not introduce traffic in the crime-scene network and generally do not modify the data that is being extracted. This process helps establish a chain of custody, a list that maintains identifiers for individuals that have had access to the collected data, ensuring that the investigation's findings will be admissible in a court of law.

- **Collection** is the stage where the specially configured hardware, software, or hybrid tools that forensic investigators have pre-selected, usually during the identification or a preparation stage that precedes it, are employed to extract data that may be relevant to the investigation. In this stage, experts may need to physically remove electronic devices from the crime scene or create forensic images of the devices directly. In the context of NF, experts may need to collect packets from the network, store them in pcap files, or access network entities (PCs, laptops, IoT devices, routers, smartphones, etc.) to extract logs.

- **Examination and analysis** are the stages during which the collected data is processed by experts and a range of automated tools to identify data of interest, known as traces. After traces have been detected in the examination, experts will analyse them in the context of the investigation, considering the security incidents that contributed to crime (or cyber-incident), the range of affected devices, and attempting to answer five questions: who, what, where, when, and why. This refers to the culprit's identity, what corresponds to the target of the attack, where it specifies the location(s) where the incident took place, when it establishes a timeline, and why the criminal's motive is.

- **Presentation** is the final stage of the investigation process. In this stage, experts compile a report that details the process they followed and the extracted evidence. These reports may then be employed in a court of law to establish guilt.

13.4.1 Practical TCP/IP Analysis

In the Ubuntu Server VM, to launch Wireshark, do the following. Run the ifconfig command and identify the interface corresponding to the Ubuntu Server VM's IP address, as shown in Figure 13.6. Then, identify the correct interface, open a terminal and write "sudo Wireshark" to start Wireshark with root privileges. This will enable viewing all available interfaces. Finally, select the interface that was identified with ifconfig and then click Start.

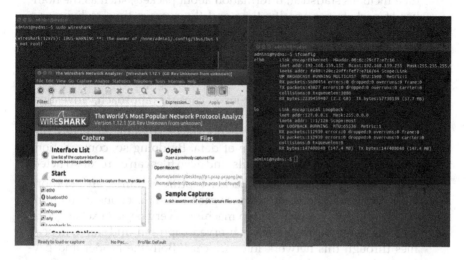

FIGURE 13.6
Using Wireshark and ifconfig to define IP addresses of machines.

13.5 Levels of Network Traffic Capture for Forensics Analysis

NF deals with security incidents that either took place inside the network or employed to establish a connection with the target system and compromise it. Thus, NF experts need to be proficient in collecting data from the network. There are various approaches experts can follow to collect network data for an investigation, which range from primary "flow data" to full packet capture and even beyond.

- **Flow data**: In flow data capture, packets are captured and transformed in a summarised format that describes the connection statistics. Each flow is uniquely identified using a quintuplet of features, known as flow identifiers, source destination port numbers/IP addresses, and the network protocol, obtained from the various headers at different TCP/IP layers. A temporal feature avoids confusion between network flows obtained between the same hosts using the same port numbers and communication protocols. Usually, a timestamp of the first packet is employed as the sixth flow identifier. Tools that can extract network flows from network communications include Argus and Bro (formerly known as Zeek).

- **Transaction data**: By capturing transaction data that could span multiple packets within a connection, experts can extract deeper connection-level information about the monitored traffic, leading to further insights about conditions during the cyberattack. Transaction data includes statistical information about packets, such as the number of packets per specific time and the average time to live per a particular time. It can be thought of as an extension of the flow data capture. Experts can employ the ZEEK tool (previously known as Bro) to gain protocol-specific logs either live from the wire or by analysing offline Packet Capture (pcap) files, such as HTTP, FTP, DNS, and SSL.

- **Reassembly data**: A packet-level capture approach, investigators employ assembly data capture methods when they require to reassemble IP packets into transport-layer streams or reconstruct the application layer exchange of data. By doing so, conversations between users exchanged emails, and even files can be reconstructed.

- **Full packet capture**: A full packet capture requires an appropriately configured network interface card in promiscuous mode to be attached to an investigator's tap machine. Every packet exchanged in the Local Area Network can be observed, captured, and stored in pcap files through this network interface card. Both the header and payload information of packets are stored, which can be used for offline analysis such as deep packet inspection. Tools used for full packet capture include NetworkMiner, netsniff-ng, and tcpdump tools.

- **Alert data:** Alert data is generated when a fine-tuned, often signature-based tool is employed to detect suspicious traffic in the network. Rules such as If source IP=destination IP source port=destination port, alert Land Denial-of-Service (DoS) attack are employed by network intrusion detection systems to generate alerts and invoke other cyber-defence or NF mechanisms. Snort and Suricata tools are two popular tools that employ such rules for the early detection and prevention of cyberattacks.

13.6 NetworkMiner Tool for Network Forensics

NetworkMiner [4] is an open-source Network Forensic Analysis Tool for Windows and Linux/Mac that primarily provides passive network sniffing/packet capturing capabilities. Operating in passive mode, NetworkMiner can detect the Operating System of hosts by analysing their generated network traffic, identifying sessions, hostnames, open ports without injecting traffic to the network. Furthermore, an offline analysis mode is provided, where the tool can access pre-collected traffic in pcap files to extract passwords, sessions, certificate files, and or packet payloads.

13.6.1 Applying the Network Forensic Investigation Process

Each step of the process shown in Figure 13.5 defines actions that need to be carried out during an NF investigation. This ensures that relevant data is collected from appropriate sources in a forensically sound manner and that the integrity of the findings cannot be questioned in a court of law. In the identification, artefacts such as network traffic and protocol patterns are automatically assessed to find indications of cyberattacks. In the preservation stage, hash digests of the collected data are generated to ensure the integrity of collected data and extracted evidence. This is an ongoing process while traffic from the suspected network is passively collected. In the analysis stage, the contents of the collected pcap files are explored, for example, using NetworkMiner, while in the Presentation stage, evidence is extracted and compiled into a report.

13.6.2 Examples of Network Forensic Investigation

Some practical examples are presented here for investigating network incidents using Wireshark and NetworkMiner. In the first scenario, we first send an image file from the Kali Linux to the Ubuntu Server VMs, and we

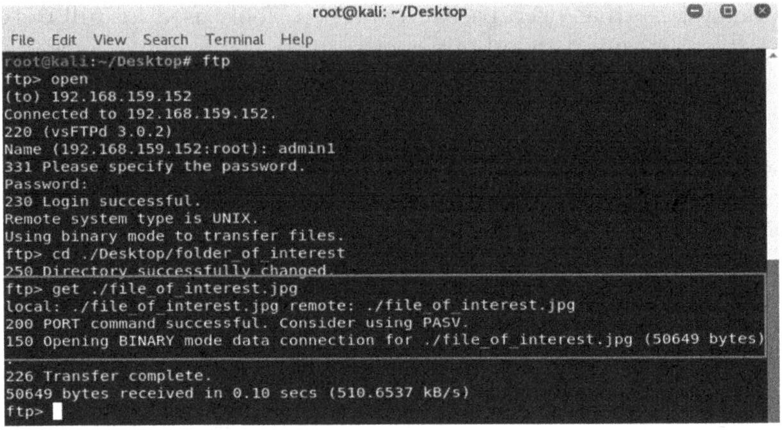

FIGURE 13.7
FTP download of a file to Kali VM.

employ NetworkMiner to intercept and reconstruct the exchanged file. At first, we employ the simple FTP command to authenticate as a legitimate user (admin1) in the Ubuntu Server VM and download the file_of_interest.jpg file to the Kali VM be seen in Figure 13.7.

In the Wireshark output, observe the collected traffic of FTP. It assists in observing any suspicious patterns (e.g. transport layer flags, number of senders, number of packets per sender), as shown in Figure 13.8.

Before the FTP transfer is initiated, the NetworkMiner tool is activated in the Windows VM. NetworkMiner automatically calculated a hash digest value for the pcap file it created during collection to ensure the integrity of our findings. NetworkMiner has detected a file exchanged through FTP among the collected packets between the Kali and Ubuntu VMs (identified by their IP addresses). As can be seen in Figure 13.9, we were able to extract the file, which can then be accessed because it was not encrypted before transferring.

The following scenario involves the detection of an SYN DoS attack. An SYN DoS attack exploits the three-way handshake, a procedure carried out prior to any TCP connection. The two hosts negotiate their communication parameters (including send and receive sequence numbers), and a sequence of synchronisation, synchronisation acknowledgement, and acknowledgement packets are exchanged. The attacker spawns multiple synchronisation packets that emulate legitimate network behaviour. The attacker sends back an acknowledgement to the initial synchronisation packet and awaits the sender to finalize the handshake, which never happens, leaving ports half-open. Using the hpin3 command in Kali, the Ubuntu Server VM was targeted, as can be seen in Figure 13.10.

While the SYN DoS flood attack is ongoing, its effects are visible in the Ubuntu server VM, as its resource consumption increases (network, CPU, memory) and in the Windows VM, as shown in Figure 13.11. It was noticed

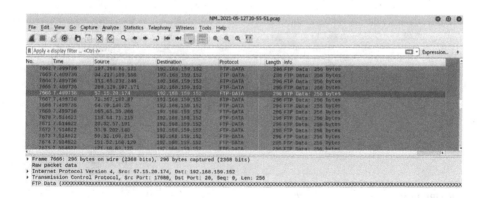

FIGURE 13.8
Observing FTP packets using Wireshark.

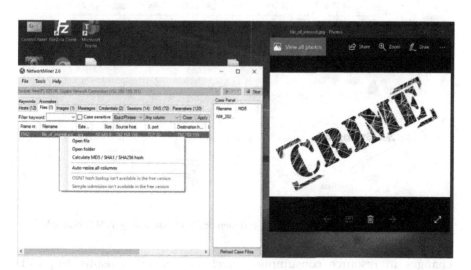

FIGURE 13.9
NetworkMiner identifying and extracting a file from network traffic.

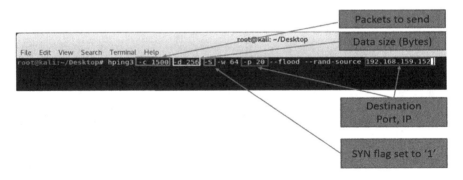

FIGURE 13.10
SYN DoS using hping3.

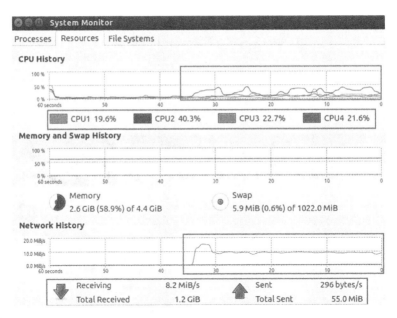

FIGURE 13.11
Consumption increases of network, CPU, and memory while launching SYN DoS attack.

changes in resource consumption, such as network streams and CPU utilization.

Figures 13.12 and 13.13 illustrate indications that point out to an attack taking place. First, multiple hosts are targeting a single IP address (that of the Ubuntu Server), and each host has sent 1 packet. Furthermore, a unique pattern of each distinct host utilizing ascending source port number is an additional indication of an attack. To verify our findings of an SYN DoS attack, we can use Wireshark and see if the exchanged packets are SYN packets.

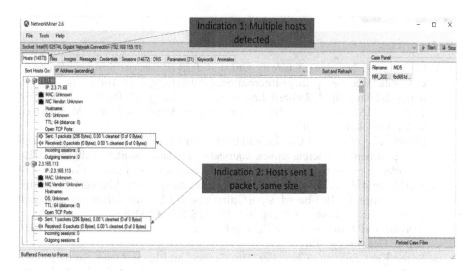

FIGURE 13.12
NetworkMiner indications 1 and 2 of an SYN DoS attack.

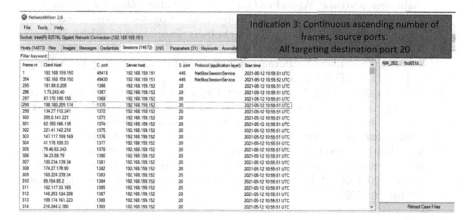

FIGURE 13.13
NetworkMiner indication 3 of an SYN DoS attack.

13.7 Conclusion

This chapter has discussed the components of NF investigations. Additionally, the OSI and TCP/IP layers were discussed to illustrate the basics of networking. Finally, practical applications of NF principles in realistic scenarios using popular tools, such as Wireshark and NetworkMiner, are discussed. Next, advanced machine learning techniques for forensics will be demonstrated.

References

1. Mohammed M Alani. *Guide to OSI and TCP/IP Models*. Springer Briefs in Computer Science. Springer International Publishing, New York, 2014.
2. James Edwards and Richard Bramante. *Networking Self-Teaching Guide: OSI, TCP/IP, LANs, MANs, WANs, Implementation, Management, and Maintenance.* Wiley, Hoboken, NJ, 2015.
3. David Goad, Andrew T Collins, and Uri Gal. Privacy and the Internet of Things-an experiment in discrete choice. *Information & Management*, 58(2):103292, 2021.
4. Nickolaos Koroniotis, Nour Moustafa, Francesco Schiliro, Praveen Gauravaram, and Helge Janicke. The SAir-IIoT cyber testbed as a service: A Novel Cybertwins Architecture in IIoT-based smart airports. *IEEE Transactions on Intelligent Transportation Systems*, 2021. doi: 10.1109/TITS.2021.3106378.
5. R Joshi and Emmanuel S Pilli. *Fundamentals of Network Forensics*. Springer, Berlin, Heidelberg, 2016.
6. Karen Kent, Suzanne Chevalier, Tim Grance, and Hung Dang. Guide to integrating forensic techniques into incident response. *NIST Special Publication*, 10(14):800–86, 2006.
7. Suleman Khan, Abdullah Gani, Ainuddin Wahid Abdul Wahab, Muhammad Shiraz, and Iftikhar Ahmad. Network forensics: Review, taxonomy, and open challenges. *Journal of Network and Computer Applications*, 66:214–235, 2016.
8. Constantinos Kolias, Georgios Kambourakis, Angelos Stavrou, and Jeffrey Voas. Ddos in the iot: Mirai and other botnets. *Computer*, 50(7):80–84, 2017.
9. Nickolaos Koroniotis, Nour Moustafa, and Elena Sitnikova. Forensics and deep learning mechanisms for botnets in Internet of Things: A survey of challenges and solutions. *IEEE Access*, 7:61764–61785, 2019.
10. Nickolaos Koroniotis, Nour Moustafa, Elena Sitnikova, and Jill Slay. Towards developing network forensic mechanism for botnet activities in the iot based on machine learning techniques. In *International Conference on Mobile Networks and Management*, Melbourne, VIC, Australia, pp. 30–44. Springer, 2017.
11. Mirosław Kutyłowski, Jun Zhang, and Chao Chen. *Network and System Security: 14th International Conference, NSS 2020*, Melbourne, VIC, Australia, November 25–27, 2020, LNCS sublibrary: Security and cryptology. Springer International Publishing, 2020.

14

Machine Learning Trends for Digital Forensics

14.1 Introduction

The previous chapter discussed network forensics, introducing networking fundamentals and the network forensic process, discussing challenges, and reinforcing the presented concepts through practical scenarios. This chapter introduces machine learning (ML) methods and their application in digital forensics (DF) scenarios. The automation through ML is explained by addressing some challenges in DF. Next, we discuss artificial intelligence (AI), including ML and deep learning (DL) methods, and DF applicability. Finally, some practical scenarios of applying ML for DF are provided, utilizing established and real datasets.

The main objectives of this chapter are as follows:
- Discussing ML for DF
- Learning ML process
- Demonstrating ML tasks and types
- Explaining applications of ML for DF

14.2 Why Do We Need Artificial Intelligence in Digital Forensics?

Recall that in a DF investigation, devices of interest that may be sources of interest to the investigation data are identified during the first three stages. They are also accessed and, through secure methods, either extracted themselves and imaged, and data is copied in a forensically sound manner. Collected data and device images then need to be transported to a forensics environment, where they will be assessed, examined, and analysed so that traces of incidents can be detected.

During this process of examination and analysis, vast data collections, consisting of heterogeneous data (logs, emails, images, pcaps, memory dumps, etc.) need to be carefully evaluated by investigators to separate benign data from potential traces of security incidents [11]. In essence, this is a repetitive process that requires careful assessment of multiple and varied records to detect obscure patterns in the data that, in turn, may lead to the identification of traces.

Challenges arise for investigators when tasked with such actions. Due to human error, they may be incapable of identifying these patterns or overlook critical records in the vast data collection. This, coupled with the implied requirement for an investigation to be completed in a timely fashion so that any findings can be presented in a trial, leads to the need for automated tools that can assist forensic experts during the examination and analysis stages. With these requirements in mind and a clear motivation for transitioning towards automated methods that may accelerate the DF investigation process, researchers have started incorporating AI in DF tools [9].

14.2.1 Artificial Intelligence for Digital Forensics

As AI, we can consider any system that has been programmed, trained, or otherwise prepared to resemble and employ human-like characteristics. This can assist in problem-solving, often beneficially utilizing its environment and actively attempting to solve a given task and achieve a goal. AI can take many forms and be employed in various settings, from simple recommendation systems and auto-correction to identifying gravitational waves in space and automatically generating music [4,5]. Although AI consists of multiple sub-disciplines, each seeks to fulfil the goals of AI differently. The most popular AI techniques that have emerged over the years are ML and DL.

In ML, statistical methods, algorithms, and advanced mathematics are utilized to allow machines to learn from data, either with or without human intervention, and gradually improve without being explicitly programmed. Models such as decision trees, Naive Bayes, and neural networks are fed data in multiple iterations and progressively "learn," enabling them to make predictions from past experiences, group data based on similarities, or generate new data. Figure 14.1 provides a Venn diagram illustrating ML components, including data sources and features extracted from the data to enhance predictive capabilities and algorithms that dictate how learning occurs.

Of particular interest from these ML models is the neural network in all its permutations (recurrent neural network, convolution neural network, generative adversarial networks, etc.). This interest has led to further development, and through focused research, a new ML sub-discipline emerged, DL.

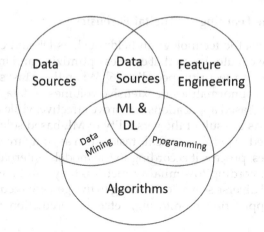

FIGURE 14.1
Venn diagram of ML components.

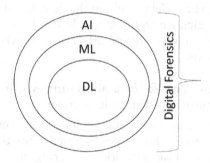

FIGURE 14.2
Hierarchy of AI, ML, and DL with DF.

Essentially, DL is a sub-discipline of ML and refers to neural networks that have been designed with a very "deep" architecture, either having multiple (in the hundreds) neurons in each layer, multiple layers, or both. Through experimentation and empirical assessment, it has been observed that DL models display increased performance when tasked faced with vast quantities of data.

Some DL types, such as recurrent neural networks, can employ a form of memory to improve future predictions, regulating this memory according to performance, without requiring excessive time during execution. The high performance of ML and DL models, as indicated by the research community's attention and the produced work, render those prime choices for addressing the automation requirements set in the DF domain. The hierarchy that was discussed between AI, ML, and DL for their use in DF can be seen in Figure 14.2.

14.2.2 Machine Learning for Digital Forensics

ML is a very versatile technology in fields such as DF and cybersecurity. It can provide power automation that enables optimizing and informing forensic investigators and security specialists [6]. When these tasks involve detecting patterns and abnormalities in extensive volumes of data, the automation provided by ML-based applications can prove effective while accelerating the investigations. As a result of the versatility of ML-based solutions, although they are designed to perform specific tasks, such as compare inbound data to known instances, process it according to the models' internal state or group data together according to similarity metrics, they can be employed effectively in several diverse scenarios, leading to five general security categories/modes of ML application: monitoring, detection, prediction, prevention, and response.

Through constant monitoring, cyber threats can be detected by analysing inbound traffic and its statistics, writing requests on a hard drive or the state of CPU registers in passive mode. If a threat is detected, response measures are invoked, where short-term remediation actions are sought, such as blocking IPs that appear to be collaborating in the detected attack. The prediction and prevention activities are gathered in a low-risk timeframe, with predictive actions seeking to foresee future attack trends, which will help define the prevention methods.

By considering the process of assimilating ML in digital forensic activities, three questions/dimensions arise:

- **Why** is the first dimension and refers to the scenarios and goals for which ML models have been selected. This dimension defines whether ML is used for network monitoring, threat detection, malware analysis, and network attack tracing.
- **What** is the second dimension and corresponds to the technical level at which ML is applied in the selected scenarios. This dimension defines whether ML will be used for cybersecurity monitoring purposes, such as in an intrusion detection system or for DF to detect traces in data collections.
- **How** is the third dimension and addresses how data will be processed by the ML application in the DF scenarios. For example, will models process data in transit (e.g. transmitted through a network) or at rest, online, or offline?

These three dimensions help place ML in perspective, establishing the purpose (why) and method (what) of the application and the type of data (how) that will be processed. Consider the scenario in which a cybersecurity incident has occurred, and ML will be applied for a *post-facto* investigation. In which of the six phases of the DF investigation process would ML be most

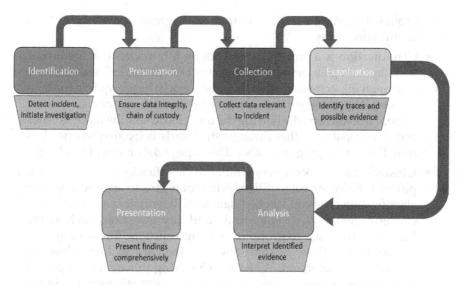

FIGURE 14.3
Digital forensic investigation process and ML.

appropriately used? An argument could be made for any phase. However, the ability of ML models to process thousands of records swiftly would suggest that the examination and analysis phases would benefit the most, as shown in Figure 14.3.

14.2.3 Machine Learning Basics

The field of ML consists of a variety of various algorithms and models, each employing data to build knowledge internally. This extracts knowledge and patterns and correlates records based on similarity, resulting in several common tasks/categories within which any ML model can be placed [12].

- **Dimensionality reduction** is a form of feature reduction or feature selection that employs for ML to clean and filter data. It is considered a wrapper method. It selects features by repeatedly training an ML model on multiple subsets derived from the original feature space and evaluating them (the models). This is by using metrics such as accuracy, precision, recall, etc., as explained in Chapter 2. The feature subset that produced the best model is the one that is selected.

- **Regression** is a process carried out by ML models tasked with predicting future values, according to previous observations. A regressor can be thought of as a model that attempts to approximate the behaviour of a mathematical equation, which it builds based on the provided training data. The output of a regression model is often a continuous value.

- **Classification** is a process carried out by ML models that are tasked with separating observations into distinct categories, known as classes, by employing knowledge that they (the models) have acquired through training. The training data is generated so that at least one class value is present in each record, allowing the model to assess the input features and derive patterns that distinguish records belonging to one class from those belonging to another. This type of data is called labelling.

- **Clustering** is a process very similar to classification in their general purpose; however, each utilizes different means to achieve it. Where classification relies on labelled data to differentiate between classes, clustering employs distance and similarity measures, such as the Euclidean distance, to group records into clusters. The user defines the number of these clusters; however, the clusters themselves are not known beforehand. When the clustering process is completed, records in each cluster should be very similar to other records in the same cluster and dissimilar to records in other clusters.

- **Association rule mining** is a process involving the generation of recommendations and rules, based on patterns in the data. Values of features are assessed, and measures such as frequency of occurrence and conditional probabilities are employed to identify combinations that appear most often.

- **Generative models** are ML models that rely on internalizing the distribution of data and employing it to generate new, often novel data.

14.3 Machine Learning Process

Building an effective ML model involves a series of steps, each of which is a prerequisite of the next and serves a specific purpose, as shown in Figure 14.4.

With a particular scenario in mind, thus specifying the intended purpose and setting in which the ML model will be trained, the first step is to select the category of ML model (as they were presented in the previous section. It is important to consider the problem we are trying to solve first, and specify the ML model category, such as clustering, classification, or regression, as this will affect the required data. Following are the common steps associated with preparing an ML model for real-world application [1,8].

14.3.1 Data Collection and Pre-Processing

The first actual step in building an ML model is data collection and pre-processing. Raw data is gathered from potentially multiple sources,

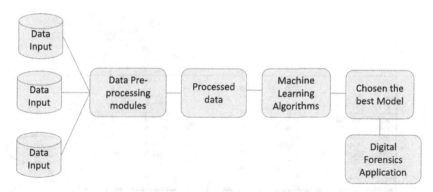

FIGURE 14.4
The process of preparing an ML model.

resulting in a collection of heterogeneous files that can include logs, network packets, images, videos, and memory dumps. Before we can utilize this data to train an ML model, it needs to be properly prepared, modified, or even augmented, tasks that occur during pre-processing. In the pre-processing stage, files are scanned, information extracted and organized in a format suitable for the following stages; this often, but is not limited to, a tabular format.

Additional actions that may take place during pre-processing include a unification of all feature scales by re-scaling the data through normalization or standardization so that features with an outlier or very large values do not dominate the other features in the ML model. Analysis of the now modified data can now take place to gleam potential patterns that may lead to the generation of new features that could improve the performance of the ML model due to their increased expressiveness. On the other hand, if the collected data has high dimensionality and thus too many features, it could be pertinent to employ methods for feature reduction, such as wrapper methods that are dependent on a model or filter methods that rely on mathematical calculations. This iterative process persists until a user-set condition that illustrates the quality of the gathered and pre-processed data has been reached.

In the context of DF, including network forensics which was discussed in the previous chapter, recall that during an investigation, heterogeneous data may be collected, including network flow records, alert data from network and host intrusion detection systems, Syslog data, transaction data, full packet captures and more. An example of network flow data can be seen in Figure 14.5, where the flow records are either placed in the training or testing set, a portion of the former being utilized for validation. Individual rows are also known as records, vectors, or observations and are essentially individual data points that an ML model will process.

Columns are known as attributes, fields, or features, and can be thought of as the data's dimensions, with a record consisting of a single feature value for

Digital Forensics in the Era of Artificial Intelligence

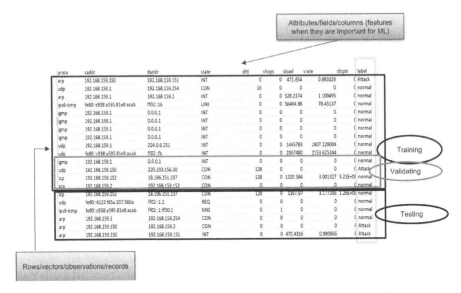

FIGURE 14.5
Example of network flow data.

each feature in the dataset. The pre-processing activity of feature generation or feature extraction involves deriving new features from the original feature space that are more expressive and do not increase redundancy. For example, as shown in Figure 14.6, the initial packets from a network capture are processed, and features are extracted to form the network flow representation before labelling occurs, where the class feature is added. In other scenarios, when processing emails or URLs, other feature extraction methods can be applied, such as the occurrence count for each word in the data.

14.3.2 Training and Testing Phases

The next step is to select a model suited to the task and prepare its training and testing sets. Recall that we have already chosen the model type (classification, clustering, etc.) to exclude models from other categories. However, this choice can change if the nature of the data and the pre-processing steps permit it. In this stage, the data needs to be separated into either 2 or 3 subgroups: 1 for training, 1 for testing, and an optional 1 for validation. As the name suggests, the training set is the portion of the data utilized to train the model. The validation set is employed to evaluate the ML model's performance during the training steps. The rest of the data forms the testing set and is used to assess the model's performance after the training, resulting in metrics such as accuracy, precision, recall, specificity, etc.

Although there is no strict rule for splitting the data, a general rule of thumb is to utilize 80%–70% for training, as we wish to employ most of our

No.	Time	Source	Destination	Protocol	Length Info
20	3.101279492	Vmware_7b:e9:a0	ubuntu.mydns.com	ARP	42 192.168.159.151 is at 00:0c:29:7b:e...
21	3.108066582	mobiledb.mydns.com	Broadcast	ARP	60 Who has 192.168.159.2? Tell 192.168...
22	3.316889624	Vmware_7b:e9:a0	ubuntu.mydns.com	ARP	42 Who has 192.168.159.152? Tell 192.1...
23	3.317387463	ubuntu.mydns.com	Vmware_7b:e9:a0	ARP	60 192.168.159.152 is at 00:0c:29:e2:a...
24	3.732912393	ubuntu.mydns.com	Broadcast	ARP	60 Who has 192.168.159.2? Tell 192.168...
25	4.132119447	mobiledb.mydns.com	Broadcast	ARP	60 Who has 192.168.159.2? Tell 192.168...
26	5.155731939	mobiledb.mydns.com	Broadcast	ARP	60 Who has 192.168.159.2? Tell 192.168...
27	6.180137129	mobiledb.mydns.com	Broadcast	ARP	60 Who has 192.168.159.2? Tell 192.168...
28	6.734981424	ubuntu.mydns.com	Broadcast	ARP	60 Who has 192.168.159.2? Tell 192.168...
29	7.204005376	mobiledb.mydns.com	Broadcast	ARP	60 Who has 192.168.159.2? Tell 192.168...
30	7.733053790	ubuntu.mydns.com	Broadcast	ARP	60 Who has 192.168.159.2? Tell 192.168...
31	8.091361051	ubuntu.mydns.com	192.168.159.151	DNS	81 Standard query response 0x23e4 Serv...
32	8.091390198	ubuntu.mydns.com	192.168.159.151	DNS	81 Standard query response 0x16a3 Serv...
33	8.091522812	192.168.159.151	ubuntu.mydns.com	DNS	91 Standard query 0x06d6 A 2.ubuntu.po...
34	8.092069958	192.168.159.151	ubuntu.mydns.com	DNS	91 Standard query 0xd20d AAAA 2.ubuntu...

▸ Frame 31: 81 bytes on wire (648 bits), 81 bytes captured (648 bits) on interface 0
▸ Ethernet II, Src: ubuntu.mydns.com (00:0c:29:e2:a8:ad), Dst: Vmware_7b:e9:a0 (00:0c:29:7b:e9:a0)
▸ Internet Protocol Version 4, Src: ubuntu.mydns.com (192.168.159.152), Dst: 192.168.159.151 (192.168.159.151)
▸ User Datagram Protocol, Src Port: domain (53), Dst Port: 51010 (51010)
▸ Domain Name System (response)

Pcap files

Extracting Features and data label

state	dttl	shops	sload	srate	dtcpb	label
INT	0	0	471.854	0.983029	0	Attack
CON	16	0	0	0	0	normal
INT	0	0	528.2374	1.100495	0	normal
UNK	0	0	56484.98	78.45137	0	normal
INT	0	0	0	0	0	normal

FIGURE 14.6
Feature extraction and labelling on network data.

data for teaching the model and 20%–30% for testing. It should be stated that the testing set should not include records present in the training set, as the purpose of the testing phase is to evaluate the performance of the trained model on previously unseen data. By doing so, we can deduce the model's effectiveness and make adjustments by tweaking the hyperparameters that affect a model's performance, such as the learning rate, batch size, and the number of iterations. Essential to this assessment process of a trained model is calculating specific metrics derived from the correct and incorrect processing of data.

For example, in binary classification, although applicable to multi-class classification [8], a structure depicts the outcome of the testing phase of an ML model, known as the confusion matrix. It provides four core values, true positive, true negative, false positive, and false negative, with the former 2 depicting the records correctly classified as positive and negative respectively, while the latter 2 representing the incorrectly classified as positive (thus originally negative) and incorrectly classified as negative (thus originally positive) records respectively. Note that these four values can be used to calculate metrics such as accuracy; however, it should not rely primarily on the accuracy, as it can be affected by class imbalances.

14.4 Applications of Machine Learning Models

The final stage is reached after the ML model has been optimized by selecting the most appropriate model and then altering its hyperparameters. The now finalized model is prepared and can be effectively utilized in real-world data and scenarios.

14.4.1 Machine Learning Types

There are various ways to group ML models; for example, depending on how they process data, a model can be considered a classifier if it assigns values to the class feature of records, or a clustering model if it creates record groups depending on similarity. Another way of grouping ML models would be how humans interact with them during training, resulting in four main types: supervised, unsupervised, semi-supervised, and reinforcement learning [3,13].

- **Supervised learning** defines models that make predictions on new events by utilizing previously acquired knowledge, for example, by learning the probability distributions of previous events and observations. For this, the class (label) feature values need to be provided to the model; thus, a part of the pre-processing stage is labelling, performed by experts. Two popular ML model categories that belong to the supervised learning type are classification and regression models. Both model types involve the prediction of some value. However, classification models predict the class feature of a record, while regression is often employed to predict a continuous value, such as stock prices. Examples of regression models include linear regression, polynomial regression, and ridge regression. Examples of classification models include logistic regression, K-nearest neighbours, support vector machine, decision tree, Naive Bayes, artificial neural networks, and recurrent neural networks.

- **Unsupervised learning** defines models that attempt to find patterns in data without requiring a class (label) feature during training. To do this, unsupervised models, a popular example of which are clustering algorithms, employ similarity measures, resulting in groups (clusters) that are most similar and whose records are dissimilar between different groups. An example of an unsupervised clustering algorithm would be K-means, which starts with a random selection of centroids (mean), one for each cluster. Next, the distance of each record between all centroids is calculated, with the records assigned to the cluster of the centroid that is closest to them (using, for example, the Euclidean distance). Then, new centroids are calculated for each cluster, and the process repeats until the centroids remain relatively unchanged.

- **Semi-supervised learning** tries to combine the benefits of both supervised and unsupervised approaches and is often employed when the majority of the data we have is unlabelled, and there are some labelled data. The model is trained on the labelled data and then is tasked with classifying the unlabelled data. From the now classified unlabelled data, we keep the records for which we have high confidence that they have been correctly classified and add them to the labelled data, with the process repeating until there are no more unlabelled data.

- **Reinforcement learning** is an environment-driven approach that can be used when the model needs to respond/react to the changes in its environment. It's like a kid who is learning environment by trial and error. The model starts by taking random actions, and periodically, by receiving "rewards" from a dedicated function, it learns the correct reactions depending on the intended goal and the environment.

14.5 Case Study: Using the TON_IoT Dataset for Forensics

The quality of data directly impacts the quality of a trained ML model. Curated datasets are a requirement for developing robust ML-based applications. Although there is a lack of such datasets for DF and cybersecurity purposes, recent research efforts have sought to address this issue, with datasets such as the Bot-IoT [7] and the TON_IoT dataset [2,10]. Both datasets represent intelligent environments, where IoT devices operated and were interconnected with multiple virtual and physical devices to generate normal traffic instances. However, as the latter includes traces from four distinct sub-datasets, this case study focuses on the TON_IoT dataset.

To represent abnormal behaviour, remote cyberattacks were launched, targeting Windows and Linux machines and the IoT devices, resulting in a Windows, a Linux, a network, and an IoT sub-dataset. After the pre-processing stage, where data from the aforementioned heterogeneous sources was collected, transformed, unified, and enhanced through feature generation, the finalized TON_IoT dataset was generated. A sample of the features relating to HDD traces from the TON_IoT dataset can be seen in Figure 14.7.

Recall that network flows are summarised representations of network packets, where we maintain the network flow identifiers such as proto (protocol), saddr (source IP address), and daddr (destination IP address), as shown in Figure 14.8. These features should be removed while applying ML

ID	Feature	Type	Description
103	LogicalDisk(_Total)/Avg. Disk Bytes/Write	Number	The average number of bytes transferred to the disk during write operations.
104	LogicalDisk(_Total)/pct_ Idle Time	Number	The percentage of time during the sample interval that the disk was idle.
105	LogicalDisk(_Total)/Disk Reads/sec	Number	The rate of read operations on the disk.

FIGURE 14.7
Sample of features from hard drive traces from the TON_IoT dataset.

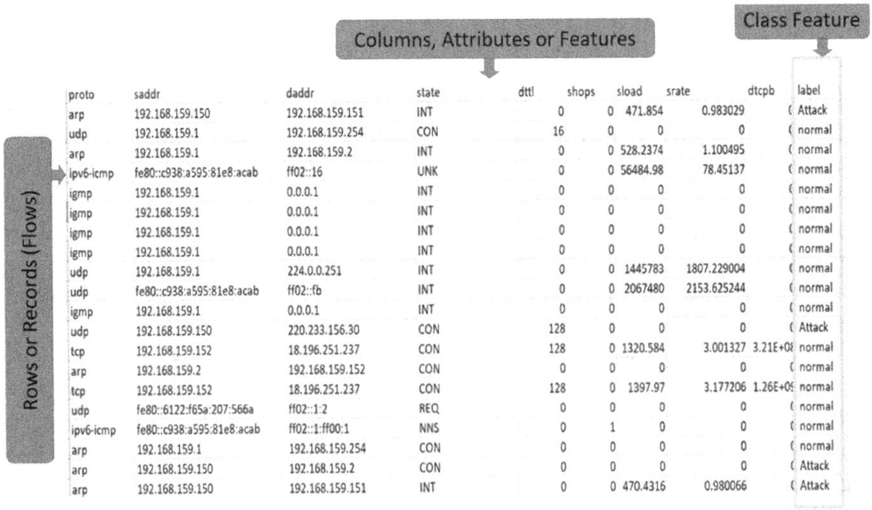

proto	saddr	daddr	state	dttl	shops	sload	srate	dtcpb	label
arp	192.168.159.150	192.168.159.151	INT	0	0	471.854	0.983029	(	Attack
udp	192.168.159.1	192.168.159.254	CON	16	0	0	0	(	normal
arp	192.168.159.1	192.168.159.2	INT	0	0	528.2374	1.100495	(	normal
ipv6-icmp	fe80::c938:a595:81e8:acab	ff02::16	UNK	0	0	56484.98	78.45137	(	normal
igmp	192.168.159.1	0.0.0.1	INT	0	0	0	0	(	normal
igmp	192.168.159.1	0.0.0.1	INT	0	0	0	0	(	normal
igmp	192.168.159.1	0.0.0.1	INT	0	0	0	0	(	normal
igmp	192.168.159.1	0.0.0.1	INT	0	0	0	0	(	normal
udp	192.168.159.1	224.0.0.251	INT	0	0	1445783	1807.229004	(	normal
udp	fe80::c938:a595:81e8:acab	ff02::fb	INT	0	0	2067480	2153.625244	(	normal
igmp	192.168.159.1	0.0.0.1	INT	0	0	0	0	(	normal
udp	192.168.159.150	220.233.156.30	CON	128	0	0	0	(	Attack
tcp	192.168.159.152	18.196.251.237	CON	128	0	1320.584	3.001327	3.21E+0	normal
arp	192.168.159.2	192.168.159.152	CON	0	0	0	0	(	normal
tcp	192.168.159.152	18.196.251.237	CON	128	0	1397.97	3.177206	1.26E+0	normal
udp	fe80::6122:f65a:207:566a	ff02::1:2	REQ	0	0	0	0	(	normal
ipv6-icmp	fe80::c938:a595:81e8:acab	ff02::1:ff00:1	NNS	0	1	0	0	(	normal
arp	192.168.159.1	192.168.159.254	CON	0	0	0	0	(	normal
arp	192.168.159.150	192.168.159.2	CON	0	0	0	0	(	Attack
arp	192.168.159.150	192.168.159.151	INT	0	0	470.4316	0.980066	(	Attack

FIGURE 14.8
Sample data features extracted from network traffic.

models as they will be unique values and do not have any patterns that will detect attack types.

In this scenario, write and execute Python code to train and evaluate an ML model, specifically a decision tree classifier, that will be tasked with classifying network flows as either benign (0) or malicious (1) traffic. In the Kali VM, Launch Spyder by typing "spyder" in a terminal. Open a new.py file, insert the following declaration of packages, and load CSV files in Python, as shown in Figure 14.9.

In Figure 14.10, a sample code for loading data from the TON_IoT dataset is presented. For convenience, a loop is coded to parse through several CSV files that are the source of the data. Several unnecessary moves are removed before pre-processing can take place. Notice the three if-then-else blocks that remove the appropriate features depending on the type of data that we are processing. For example, we eliminate most of the flow identifiers

```
import pandas as pd
import os
from sklearn.preprocessing import MinMaxScaler
from sklearn.tree import DecisionTreeClassifier
from sklearn import tree
from sklearn.preprocessing import LabelEncoder
from sklearn.model_selection import train_test_split
from sklearn import metrics
import graphviz
path_to_files="./data/"

def read_from_CSV(name):
    return pd.read_csv("./data/"+name)

if __name__ == '__main__':
    file_lst=os.listdir(path_to_files)
    print("Files in {} path: {}".format(path_to_files,file_lst))

    for file in file_lst:#Itterate through the 3 ".csv" files tht we are going to train models on.
        if not file.endswith(".csv"):#If other file type is observed, ignore it.
            continue
        if str.lower((file)).startswith("network") or str.lower((file)).startswith("linux"):
            continue
        print(file)
        fbuffer=read_from_CSV(name=file)
```

FIGURE 14.9
Python declaration packages and load CSV files.

```
if __name__ == '__main__':
    file_lst=os.listdir(path_to_files)
    print("Files in {} path: {}".format(path_to_files,file_lst))

    for file in file_lst:#Itterate through the 3 ".csv" files tht we are going to train models on.
        if not file.endswith(".csv"):#If other file type is observed, ignore it.
            continue
        if str.lower((file)).startswith("network") or str.lower((file)).startswith("linux"):
            continue
        print(file)
        fbuffer=read_from_CSV(name=file)
        print(fbuffer.head())#View the top 5 records.
        print(fbuffer.type.unique())#View unique values of a field. (In this case, the class feature attack "type")
        del fbuffer["type"]#Remove the second class feature "type", as we will train a binary classifier.
        if str.lower(file).startswith("network"):
            del fbuffer["ts"]#We will also remove the timestamp denotin the beginning of a network flow,
            del fbuffer["src_ip"]#along with the flow identifiers, as we wish for our model to learn to detect attacks
            del fbuffer["src_port"]#bas on their (the attack) characteristics, and not primarily look at the host identifiers.
            del fbuffer["dst_ip"]
            del fbuffer["dst_port"]
        elif str.lower(file).startswith("linux"):
            del fbuffer["ts"]
            del fbuffer["PID"]
        elif str.lower(file).startswith("windows"):
            del fbuffer["ts"]
```

FIGURE 14.10
Sample of Python code for loading the Ton_IoT dataset.

(timestamp, source/destination IP/port number), as the model can learn to predict attacks based on those features, ignoring the other statistical features that describe the flows.

In the pre-processing stage, as shown in Figure 14.11, we parse the features of the loaded dataset in a loop. Each feature is scanned and, if it contains only 1 unique value, it is eliminated. While if it is not in numeric format, the *LabeEncoder* is employed, which produces 1 numeric value for each unique value of the feature. For example, if the unique values are "TCP, UDP, ICMP," they are substituted by "1, 2, 3."

With the data ready, the next step is to split it into training and testing, as shown in Figure 14.12. The training set is 70% of the original in this code, with the remaining 30% employed for testing. The *DecisionTreeClassifier* from the sklearn library is then selected to train a Decision tree on the training set.

The confusion matrix is automatically generated by using Sklearn metrics. confusion_matrix method, and we can then use existing methods to calculate further the accuracy, precision, recall. It is vital to note that higher values of these metrics imply a high-performing ML model. More evaluation metrics can be used to ensure that your model has not learned to exploit

```
for fture in fbuffer.columns:
    if len(fbuffer[fture].unique())<=1:#We will automaticall delete any features that have
        # 1 unique value, as they do not add any value to the model.
        del fbuffer[fture]
    else:
        dtp=str(fbuffer[fture].dtype)
        if dtp is "object": #if our features are not in numerical form
            # we will need to convert them for sklearn's decision tree to work.
            print(fbuffer[fture].unique())
            print(fbuffer[fture].dtype)
            fbuffer[fture]=fbuffer[fture].astype(str)
            fbuffer[fture]=LabelEncoder().fit_transform(fbuffer[fture])
```

FIGURE 14.11
Sample of Python code for pre-processing the TON_IoT dataset.

```
y=fbuffer["label"]
x=fbuffer.iloc[:,0:-1]#Exclude the class feature from the input data.

x_train,x_test,y_train,y_test=train_test_split(x,y,test_size=0.3,random_state=22,stratify=y)

clf=DecisionTreeClassifier().fit(x_train,y_train)
print(tree.export_text(clf))#Visualise Decision Tree in text format

y_pred=clf.predict(x_test)
print("Calculating confuon matrix")

ytest=y_test.to_numpy()
tn,fp,fn,tp=metrics.confusion_matrix(ytest,y_pred).ravel()
print("TP: {}\tTN: {}\nFP: {}\tFN: {}".format(tp,tn,fp,fn))
print("Accuracy: {}".format(metrics.accuracy_score(ytest,y_pred)))
print("Precision: {}".format(metrics.precision_score(ytest,y_pred)))
print("Recall: {}".format(metrics.recall_score(ytest,y_pred)))
```

FIGURE 14.12
Sample Python code for training and testing a decision tree on the TON_IoT dataset.

certain circumstances to falsely indicate high performance (for instance, in unbalanced data, where one class is greatly underrepresented, and the model miss-classifies it can result in high accuracy). An additional metric you can use is Specificity or True Negative Rate (you can think of specificity as the number that conveys how many negative records were correctly classified as negative, a.k.a. normal instances in our scenario. To sum up, ML can assist investigators in discovering cyberattacks and investigate the traces and origins of these attacks automatically.

14.6 Conclusion

This chapter has discussed the value of ML for DF, associating the needs of the DFIP with the services offered by ML models. The ML process, including data collection, pre-processing, training and testing phases, and evaluation of a model's performance, were also explained. Various case studies were presented to demonstrate the significance of ML for automation.

References

1. Ethem Alpaydin. *Introduction to Machine Learning*. Adaptive Computation and Machine Learning. MIT Press, Cambridge, MA, 2004.
2. Abdullah Alsaedi, Nour Moustafa, Zahir Tari, Abdun Mahmood, and Adnan Anwar. Ton_iot telemetry dataset: A new generation dataset of iot and iiot for data-driven intrusion detection systems. *IEEE Access*, 8:165130–165150, 2020.
3. Michael W. Berry, Azlinah Mohamed, and Bee Wah Yap *Supervised and Unsupervised Learning for Data Science*. Unsupervised and Semi-Supervised Learning. Springer International Publishing, New York, 2019.
4. Prafulla Dhariwal, Heewoo Jun, Christine Payne, Jong Wook Kim, Alec Radford, and Ilya Sutskever. Jukebox: A generative model for music. arXiv preprint arXiv:2005.00341, 2020.
5. EA Huerta, Asad Khan, Xiaobo Huang, Minyang Tian, Maksim Levental, Ryan Chard, Wei Wei, Maeve Heflin, Daniel S Katz, Volodymyr Kindratenko, et al. Accelerated, scalable and reproducible ai-driven gravitational wave detection. *Nature Astronomy*, 5(10):1–7, 2021.
6. Nickolaos Koroniotis, Nour Moustafa, and Elena Sitnikova. Forensics and deep learning mechanisms for botnets in Internet of Things: A survey of challenges and solutions. *IEEE Access*, 7:61764–61785, 2019.
7. Nickolaos Koroniotis, Nour Moustafa, Elena Sitnikova, and Benjamin Turnbull. Towards the development of realistic botnet dataset in the Internet of Things for network forensic analytics: Bot-iot dataset. *Future Generation Computer Systems*, 100:779–796, 2019.

8. Abhishek Mishra. *Machine Learning in the AWS Cloud: Add Intelligence to Applications with Amazon SageMaker and Amazon Rekognition.* Wiley, Hoboken, NJ, 2019.

9. Reza Montasari and Hamid Jahankhani. *Artificial Intelligence in Cyber Security: Impact and Implications: Security Challenges, Technical and Ethical Issues, Forensic Investigative Challenges.* Advanced Sciences and Technologies for Security Applications. Springer International Publishing, New York, 2021.

10. Nour Moustafa. A new distributed architecture for evaluating ai-based security systems at the edge: Network ton_iot datasets. *Sustainable Cities and Society*, 72:102994, 2021.

11. Gabriella Punziano and Angela Delli Paoli. *Handbook of Research on Advanced Research Methodologies for a Digital Society.* Advances in Knowledge Acquisition, Transfer, and Management. IGI Global, Hershey, PA, 2021.

12. Ian H Witten, Eibe Frank, and Mark A Hall. *Data Mining: Practical Machine Learning Tools and Techniques.* The Morgan Kaufmann Series in Data Management Systems. Elsevier Science, Amsterdam, 2011.

13. Alexander Zai and Brandon Brown. *Deep Reinforcement Learning in Action.* Manning Publications, Shelter Island, NY, 2020.

Index

Printed in the United States
by Baker & Taylor Publisher Services